Giuseppe Caglioti

The Dynamics of Ambiguity

With 60 Figures, Including 10 in Colour

Springer-Verlag Berlin Heidelberg Gmbh

Giuseppe Caglioti
Dipartimento di Ingegneria Nucleare
Politecnico di Milano
CESNEF
Via Ponzio 34/3, I-20133 Milano, Italy

Translator:

Anita Olaechea Bucci
Via XXII Luglio 37, I-43100 Parma, Italy

Title of the Italian original edition: *Simmetrie infrante nella scienza e nell'arte*

The cover picture is based on an artwork by
Franco Grignani (1980): Tensione iperbolica 1029, 51 × 51 cm.
For details see the same figure given as Plate VI on page 53

Photo of the author on the back of the cover by Gilberto Signifredi.

The attached *bookmark* shows a reduced version of a blow-up of the Mandelbrot set (figure on p. 188, map 42, 100, 101 in the book by H.-O. Peitgen and P. H. Richter "The Beauty of Fractals", Springer, Berlin Heidelberg New York Tokyo 1986) and demonstrates also a laser quilting of the same figure by LASIT, I-80069 Vico Equense, Italy.

Mathematics Subject Classification (1991): 00 A xx

ISBN 978-3-642-63463-5 ISBN 978-3-642-58080-2 (eBook)
DOI 10.1007/978-3-642-58080-2

Library of Congress Cataloging-in-Publication Data
Caglioti, Giuseppe.
[Simmetrie infrante nella scienza e nell'arte. English]
The dynamics of ambiguity / Giuseppe Caglioti. p. cm.
Translation of: Simmetrie infrante nella scienza e nell'arte.
Includes bibliographical references and index.

1. Symmetry. 2. Science. 3. Art. I. Title.
Q172.5.S95C3413 1992 500 - dc20 91-3462 CIP

Originally published by Springer-Verlag Berlin Heidelberg New York in 1992
Softcover reprint of the hardcover 1st edition 1992

Typesetting: Springer TEX in-house system
41/3140-543210 Printed on acid-free paper

To My Father

Preface to the English Edition

Whenever I had a chance to listen to a talk by Giuseppe Caglioti about his ideas on the profound relations between art and science, I was fascinated. Therefore I am delighted that Caglioti has exposed his thoughts in his book, first published in Italian, and now in its English translation.

Quite often, science and art are considered as the "two cultures" dividing our society into two groups. But Caglioti, who is an excellent physicist and has a deep understanding of art, shows how this gap can not only be bridged, but how important phenomena in science and art have a common root. To this end he uses concepts such as symmetry, order, entropy, and information and refers to synergetics and nonequilibrium thermodynamics. He develops the most remarkable ideas on the perception of ambiguous figures and thus establishes relations with the cognitive sciences.

This book will be of great interest to all people who are interested in art or science. I am sure that anyone starting from his own background soon will enter into resonance with Caglioti's approach, which will carry him beyond the cultural "gap".

There is one issue on which I do not entirely share Caglioti's opinion. Synergetics and nonequilibrium thermodynamics do not only have different and independent roots, but also their concepts and methods are entirely different. This comment of mine does not lower the value of Caglioti's book, however. Quite on the contrary it shows that Caglioti is dealing with "hot subjects" presently occurring in a revolution of our scientific thinking. I am sure the readers will be fascinated by this book.

Stuttgart, June 1992 *H. Haken*

Author's Preface to the English Edition

Several years have elapsed since this book was first published in Italian. And yet my views on the subject have changed only slightly. This is not surprising. In fact the book – whose original title was *"Broken Symmetries in Science and Art"* – focuses on the symmetry and the ambiguity accompanying all processes of symmetry breaking, both in nature and in culture. And symmetry – synonym of invariance – by definition ignores the flow of time.

Some changes have been introduced however, besides the obvious elimination of a few misprints and the unavoidable insertion of new ones. With the aim of making the book more readable, one of the technical appendices – originally Appendix III, on broken symmetries and spectroscopic effects at the origin of perception of colors – has been eliminated. On the other hand, three new inserts – on *exchange symmetry*, on *selfsimilarity in nature and culture: the fractals*, and on *the game of broken symmetries in the industrial trademark* – with eight new figures, have been included.

It is a pleasure to thank Professor Hermann Haken for his preface, Springer-Verlag for its interest and confidence in my work, Mrs. Anita Olaechea Bucci for her dedicated and careful work of translation and the Ministero Affari Esteri, Italy, for its financial contribution to the translation expenses.

Milano, February 29, 1992 *Giuseppe Caglioti*

as the revelation of a reality whose completeness is such as to make any aprioristic rational "foundation" in the analytic sense impossible. Obviously this doesn't mean that such a reality isn't alive with conceptual developments, but that the basic intuition of this book can be approached only a posteriori, that is by deduction, like all those moments of thought in which the human mind, expressing its dialectic capacities to the utmost, is able to grasp the coincidence or juxtaposition of contrasting aspects of reality.

Rational measurement and intuitive evaluation

In fact, in pointing out the strict analogy existing between physical observation and cultural perception, the author achieves an unusual result – that of rediscovering the affinity which unites two logico-linguistic spheres that cultural tradition usually holds to be opposing: rational measurement, on one hand, and symbolic-intuitive evaluation on the other. This is a particularly important result, especially in an era like ours, when man seems profoundly disoriented between two orders of contrasting truths: "cold" truths that affirm the existence of phenomena and "hot" truths that give them meaning, an unquantifiable value. Now as never before, in fact, the categories of quality and quantity have drawn away from each other, becoming more and more estranged: the merit of Caglioti's work consists in the first place in bringing these two hemispheres of intellectual life together again, re-opening a close dialogue between them.

Unifying factors and Kant's a-priori categories

It is no accident that what the author expresses from the very beginning of his work possesses the fascination of a great, perhaps unwitting, philosophic revisitation. In fact, when Caglioti takes pains to point out how certain "unifying factors" (symmetry, entropy, order, information, etc.), which, while usually bound to the world of physics, appear to influence perceptive-cultural dynamics and thus apply both to the emotional and rational spheres, when the author deals with this phenomenon, he seems to offer a new edition of Kant's "a-priori" categories, taken as mental parameters according to which knowledge was organized starting with the casual disorder of the sensorial stimuli. In fact, the author writes: "Unifying factors may emerge if one considers the perceptive process carefully. During this process the sensorial stimuli, disordered at first, organize themselves and then, ordered, correlate themselves in coherent schemes which then become thought. The dynamics of perception appear to involve unifying factors such as symmetry, conservation, entropy, order, information ...". It's like a confirmation of the Kantian dictate in the light of physics' most up-to-date ideas: actually, it's obvious that the function the Königsberg philosopher attributed to his "a-priori" categories is more or less identical to that which Caglioti attributes to the "unifying factors" of physics.

Rational bases of emotion and emotional substrate of rational thought

Of course it should be said, to conclude the philosophical observations regarding the method and approach of this volume – that the "crossed" interdisciplinary procedure that animates Caglioti's work (which was mentioned at the beginning of this preface) permits the author to outline the rational bases of emotion and artistic creation, on one hand, and the emotional substrate, as well as the aesthetic intuitions of scientific-rational thought, on the other. Still it is also true that if on one hand this impostation manages to "base" the subjective sphere rationally, freeing it from that quid of imponderable usually attributed to it, on the other hand it has the effect of destabilizing the legacy of objective knowledge generally attributed to scientific thought, by pointing out its emotional assumptions, which are traditionally ignored. This fact, contrary to what one might think at first, is a salutary phenomenon and not at all alarming: it's highly positive, in fact, that it should be a professional scientist to cast doubt on and – in a certain sense – invalidate the presumed "objectivity" of experimental scientific knowledge.

Scientific "objectivity"

Concerning this, it's worth pointing out (this is a reflection that the book's thesis in itself does not imply, but that may be traced to it) that if real scientific "objectivity" does exist, it resides in the dimension of theoretical invention, which can offer the intersubjective community of man a convincing hypothesis regarding the structure of the universe, and consequently the meaning of the world, a "founding" theory in the cognitive sense, though often non-functional in the practical, daily sense. On the contrary, the presumed "objectivity" of the empirical-experimental method proves to be, in reality, no more than a kind of "enlarged subjectivity" of the anthropocentric type, within which the veracity of the empirical-experimental method is effective for us men and enables us assuredly to dominate nature and the environment, subjugating them to our ends, but certainly not to know the intrinsic truth about things, what Kant called the "noumeno" (the thing in itself), in contrast to the "phenomenon".

In other words, the criterion of practical functionality, mainly pursued by experimentally-based scientific technology, has nothing in common with the "objective" (or "total") knowledge of reality, which certainly cannot occupy the empirical dimension, too limited and circumscribed to the prerogatives of human sensation.

The presumed "objectivity" of scientific knowledge, which the procedure used by Caglioti thus contributes to destabilizing, is actually an "objectivity" based on "phenomena", that is, on the mere surface of things: in this sense, in criticizing scientific objectivism, Caglioti can boast of illustrious predecessors, the most important of whom is surely Edmund Husserl, author of a phenomenological criticism of scientific-empirical knowledge with his fundamental work "The Crisis of European Sciences and Transcendental Phenomenology". Nevertheless, as already mentioned, the great critics of scientific knowledge have almost always been philosophers or epistemologists. Here, instead, we are facing a professional scientist who, rather than criticize science in its results or in its procedures, compiles a sort of psychological anamnesis of the conceptual models that govern scientific knowledge, and draws a surprising and, for some, disturbing conclusion from it: scientific knowledge does not live apart from the psychic and emotional sphere like a god who observes the world from a distance and records its doings "objectively"; on the contrary, scientific knowledge is also immersed in the world's happenings and implicitly linked to the vicissitudes of the emotional sphere.

Scientific knowledge and emotional sphere

With this we complete the philosophical observations concerning the method and approach of Giuseppe Caglioti's work, and move on to some evaluations concerning the basic nuclei of the content.

Unquestionably, one of the supporting structures (if not the main one) of the entire work is formed by the confrontation – with its wealth of conceptual consequences – between entropy and symmetry, on one hand, and information and the breaking of symmetry on the other. Actually, this confrontation seems to evoke, in an uncommon guise, the ancient and inextinguishable dialectic bond between Dionysian and Apollonian, pulsivity and structure, content and form. That is, the perennial encounter between the two fundamental dimensions of life reappears in a scientifically "updated" version: the undifferentiated mixing of matter (entropy) and the unconscious oblivion of the codes in the stasis of the indiscernible (symmetry) clashes with the form that sets and determines, snatching matter from original chaos (information), as well as the awareness of changing and becoming (breaking of symmetry). It is a real dialogue of the highest systems, which can't help reminding us of the Platonic myth of the demiurge, which created the "entropic" and primitive chaos of matter guided by the "informative" models of hyperuranian ideas, and in this way formed life, taken as the ordering of matter.

Entropy and information, Dyonisian and Apollonian

However the dialectic relationship between entropy and information, that is between accidental disorder and structural order, seems to hint at – particularly in the dynamics of perception theorized by Caglioti – another version, in a psychoanalytical vein, of this fundamental pair of opposites: the Freudian confrontation between the concept of pleasure (libido) and the concept of reality. In the world of the unconscious, in fact, the concept of pleasure aims toward disordered satisfaction, a libidinous unbridling and a confounding osmosis with the desired object, characteristics which permit us to declare that the concept of pleasure is the "entropic" component of our unconscious and psychological life. On the other hand the confounding osmosis with the desired object, which the libido aims for, involves a voluptuousness of self-sacrifice, a "cupio dissolvi" and a loss of identity (of information, that is) which up close remind us of the indiscernibility of physical "symmetry". Passing to the other side, the concept of reality represents the structural limit to the libidinous appetitions of the unconscious, a limit formed by the relationship with the laws of the external world and the moral and intellectual world: in physical terms, it's the breaking of a longed-for "symmetry" brought on by the arrival of information.

Disorder and order, libido and reality

As may be easily deduced, the link between entropy and information (as well as the implicit link between symmetry and the breaking of symmetry) permeates our entire life, not just the world of physics but the psychological, linguistic and aesthetic spheres as well: this beats the rhythm of our interior life and represents mankind's great drama. Throughout the course of history, in fact, moments in which the current taste, culture and sensitivity were attracted by the "symmetric" dream of an almost nirvanic ecstasy or by an outburst of liberating, unbridled and "entropic" drives have always alternated with moments or phases of the opposite sign, in which man was attracted by the myth of an absolute form that could, once and for all, arrange the accidental disorder of life in a sublime "informative" structure. And certainly, even in the microcosm of each single psychological equilibrium, there will always be periods in which we would gladly liquidate the problem of life by fixing it in an absolute structure that removes every uncertainty, alternating with periods when, instead, we would like to abandon ourselves to the flow of events, letting life carry us along its course, as if in a dream. Actually, both of these extremes, if carried to great lengths, hide a danger: dominion and absolute "informative" control runs the risk of crystallizing life as in a series of prearranged formulas, while – on the contrary – the lack of any "informative" structure at all involves the impossibility of exercising any control regarding the tendential disorder that is always present in our psychological life's flow of consciousness and in the incessant flow of stimuli in the life of the senses.

Ambiguity

Finally, the culminating point of Caglioti's entire work is the concept of ambiguity, taken as bivalency, as the simultaneous presence of two aspects of reality that are incompatible with each other. A structure is ambiguous when it belongs on the side of symmetrical indiscernibility while at the same time presenting a breaking of symmetry caused by an informative element, when symmetry and information both appear in it, predominating over each other by turns. And the point of ambiguity is a very fine, ineffable dividing ridge, beyond which the observer perceives the information which reveals itself, while on this side it remains hidden and imperceptible; undoubtedly the ambiguity point is also a psychological and perceptive instant, a revelation. Wittgenstein would say that it "reveals itself" without "expressing itself explicitly".

Symmetry and dream

Awareness and information

A structure is ambiguous therefore when the opposites coincide in it, when – we could say with Nietzsche – it leads us to "dream knowing that we are dreaming", since the unawareness of the dream (symmetry) merges with awareness of it (information that breaks the symmetry of the dream), without either one damaging the other. And certainly one of the deepest aspirations of contemporary man, constantly divided between the exactness of rational truths and the profoundness of myth, is precisely this: to be able to live the grace of the dream and, at the same time, the awareness of this grace, so as to fuse – Schiller would say – the enchantment of "ingenuous" poetry with the sublime but culturally nourished inspiration of "sentimental" poetry.

Milano, September 12, 1983 *Paolo Fenoglio*

Acknowledgements

My grateful thanks to Franco Grignani, master of visual communication and design, whose work, charged with inner tension, first aroused my curiosity and then led me to venture into this insidious exploration of the structure and dynamics of thought. Generous with his time, he also honored me with his friendship, going so far as to interpret, with specially devised graphic solutions, the problems of microphysics and synergetics that arose as I went along.

I am grateful to colleagues, collaborators and friends for our many discussions, and for their critical reading of whole chapters of the manuscript: here I'd like to mention in particular Emilio Gatti, who examined the entire work, P. Bisogno, C. E. Bottani, E. R. Caianiello, S. Carra and L. Zanzi. And last but not least, my brother Luciano for his encouragement; my son Vincenzo for his scepticism, often a priori but not always unfounded; as well as my father Vincenzo for his severe but incisive and constructive criticism.

The rigorous preliminary scrutiny helped to reduce – not to zero however – the number of surviving lapses. For these, obviously, I must personally take the entire responsibility.

Warm and heartfelt thanks to Carla Cattaneo, who with constancy, precision and reliability supervised the editing of typescript and bibliography.

It would have been impossible to complete this study if the Honda Foundation, Tokyo, had not been able to include the topics presented here in the program of the International DISCOVERIES Symposiums of Rome (1977), Paris (1978) and St. Paul de Vence (1981).

I'd also like to thank the Maria Giussani Bernasconi Foundation, Varese, for having contributed to the publication costs of this research.

Giuseppe Caglioti

Table of Contents

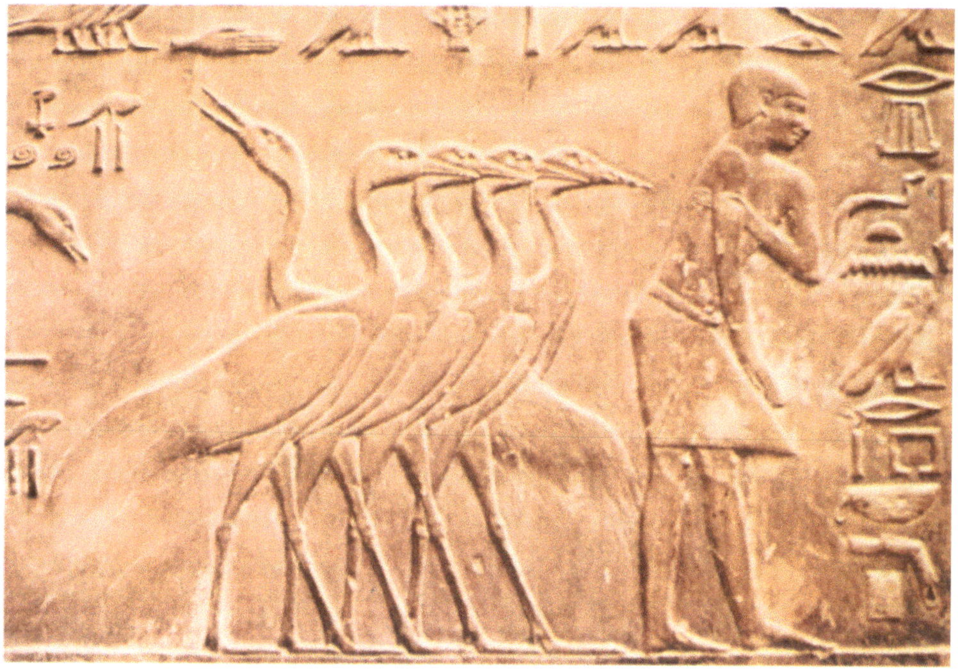

Broken Symmetry and Liveliness of the Act

Saqqara; 2500 B.C. Detail of a bas-relief from the mastaba of the Egyptian prince Ptah-hotep

Introduction

> When forms that should fly are weighty, and weigh on us psychologically; when the form-surprise substitutes the form-image (and more or less this is the case for too much baroque art, even when it's of value); when form becomes predominantly technique and a search for stylization, then baroque is no longer an historical style nor the constant factor of taste: it is a spoiled mannerism that can never be completely rehabilitated.
>
> Eugenio Montale, 1945

"We can account for everything we have measured, within a few percent." This is the closing comment of the general report on gravitational wave research, given by Edoardo Amaldi during the opening cerimony of the congress organized at Villa Olmo by the Italian Physical Society to celebrate the one hundred fiftieth anniversary of the death of Alessandro Volta. These words define quite clearly the degree of certainty in rationalizing and comprehending physical phenomena that the Galileian experimental method leads to.

Nevertheless, research laboratories apart, in the daily relationship with our fellow-men and natural structures, only rarely are those limits of rigor and certainty reached which the physicists are so proud of. Instead, now as spectator, then as protagonist, and sometimes as spectator and protagonist at the same time, each of us experiences events which follow an unpredictable and uncertain course. The few cases in which, outside the field of science, it is possible to conclude with peremptory statements like the one above, for the most part have to do with situations that are practically without interest. Our need to understand and our desire to foresee unite us. But they usually remain unsatisfied.

Man and natural structures

The relationship between man and natural structures is dominated by thought. Experienced by men since prehistoric times, this relationship involves us completely. It is so complex however, that in analyzing it one risks lapsing into the obvious; or being included, as Galileo would say "among those who write what they don't understand, and whose writings however can't be understood". Each of us, following his own nature or consistently with his own development, interprets this relationship in his own way: at times exalting its religious, artistic, poetic and sentimental aspects emotionally, at times projecting philosophic, political, social and scientific components into it rationally.

Unifying factors

The query thus arises: is it possible, taking advantage of present scientific knowledge, to identify some *unifying factor* that allows us to develop an understanding of the cultural relationship between man and natural structures, in which both the emotional aspects and the rational components inherent in the relationship itself converge?

The purpose of this study is to attempt to answer this question. We can mention some of the results in advance.

If one studies the perceptive process, unifying factors can emerge. During this process the sensorial stimuli, disordered at first, become organized, correlating themselves in orderly fashion in coherent schemes which then become thought. The dynamics of perception apparently involves unifying factors such as *symmetry*, *conservation*, *entropy*, *order* and *information*.

Symmetry, conservation, entropy, order, information

Well, the same unifying factors come under the domain of physics. They participate in fact in both the experimental observation of the structure and properties of natural systems and in the classification and description of the structural transformations that these systems may undergo while evolving.

Thus analogies seem to appear between the perceptive process on one hand, and the experimental observation and evolution of natural chemical-physical systems on the other.

I became aware of these analogies several years ago while visiting Franco Grignani's show dedicated to a "Methodology of Vision".

Perception

Being involved in my work in spectroscopic studies on the dynamics of crystalline structures, I was struck by the reproducibility and periodicity of the perceptive process of some of his "psychoplastic structures", and at the same time by the violent sensation that accompanies the perception of these structures: observe Plate I, for example[1]. A considerable portion of this study is dedicated to analyzing the "simplest" of these unrealizable structures. My curiosity aroused by the analogies between their perception and the dynamic behavior of certain well-known physical-chemical systems, I finally decided to confront the subject. The methods used in spectroscopy and synergetics turned out to be quite useful in this study. (see Insert to Chapter One).

As we were saying, thought – and, before that, perception – play a major role in the relationship between man and natural structures.

The disorder ⟶ order transformation, which during the perceptive process leads the sensorial stimuli to become thought, is in some ways similar to the second order phase transformations of chemical-physical systems and to

[1] Note that it takes several minutes to complete the entire perceptive cycle of this figure. At first it appears to be a bidimensional geometric design. Still, if the observer concentrates long enough to correlate its individual structural elements, at a certain point, abruptly, they will seem to shift, organized into a kind of three-dimensional lattice which appears to emit an intense burst of light. Any effort to fix this three-dimensional scheme in mind for more than a few tenths of a second is useless: the coherent burst of light – which we could call photophania – seems to vanish as abruptly as it appeared, and – still in the observer's mind – be replaced by the original bidimensional scheme. Following another prolonged effort at concentration and control, the cycle is repeated.

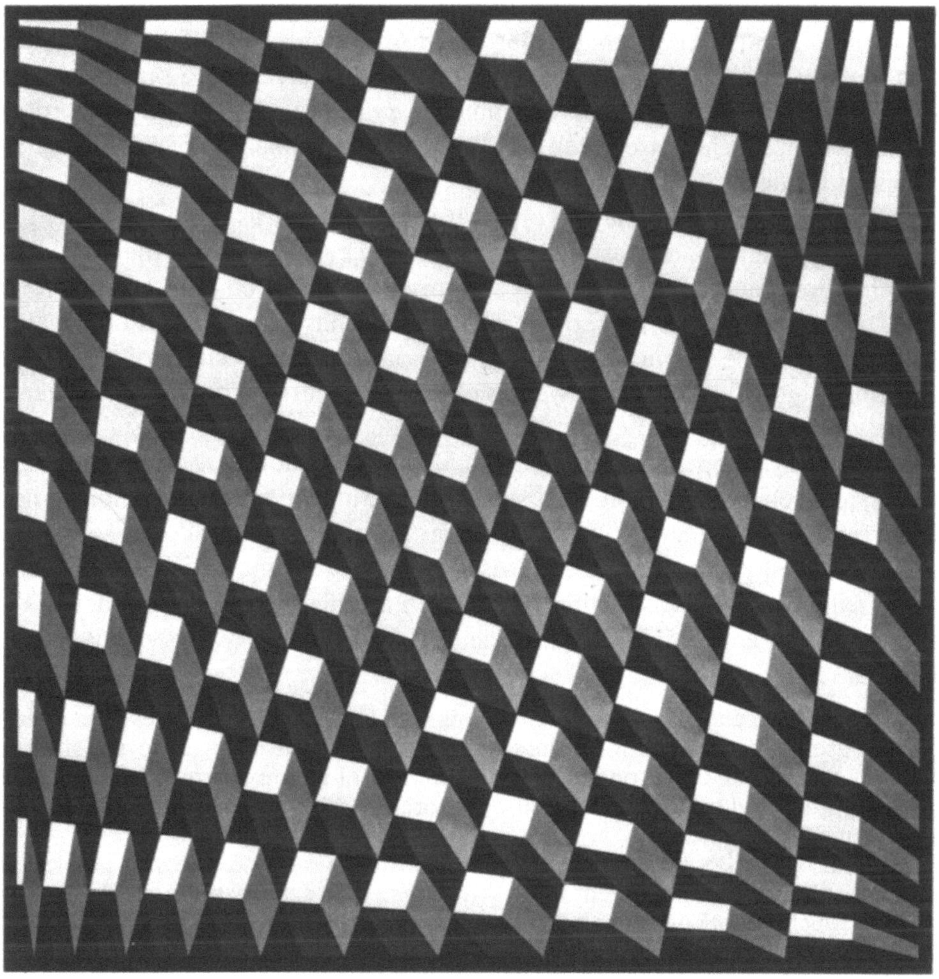

Plate I. Plastic permutation in the field, $70 \times 70\,cm^2$. Franco Grignani (1959)

After a few minutes of observation this bidimensional geometric design, interiorized in the observer's mind, seems to experience a dynamic instability. The individual graphic elements, correlated to each other during the first phase of the perceptive process, compose themselves before the observer's eyes and abruptly shift, organizing themselves in a kind of three-dimensional lattice which emanates an intense burst of light. The persistence of this three-dimensional image is fleeting however. After a few tenths of a second, just as abruptly, a reverse permutation takes place, and the inevitable relaxation of the effort of observation causes the original bidimensional scheme to reappear. To trigger the dynamic instability (or the "plastic permutation") one must exert adequate *control* over the sensorial stimuli: in other words, there must be an intense, prolonged effort of *observation, finalized* to *storing* in the memory, in the form of coherent schemes, the sensorial stimuli kindled in the mind by the interiorized design. This effort however may not be strong enough to counter the tendency of sensorial stimuli to slip one's mind. In that case, one may subject the figure to rapid, alternating rotations of about $\pm 90°$ around a horizontal axis parallel to the line joining the pupils, and look at the figure sideways, then frontally, then sideways again, and so on. Should this have no effect, the observer might try training himself first on the simpler designs in the book, and return to this panel afterwards the dynamic instabilities of synergetics. These analogies make a vivid impression on us adults (for we have long since forgotten how difficult perception was in early childhood), at the precise moment of perceiving a particular class of structures, composed of centersymmetric *ambiguous figures*. The perception of these figures lends itself well to an introspective analysis of the formation of visual thought.

Ambiguous figures

Since we are emotionally involved, it is possible for us, in dynamically perceiving these figures, to grasp the essence of the basic processes that lead from the chaotic tumult of the sensorial stimuli to the onset of ordered thought.

The analysis of the perceptive mechanisms that lead us from observation to thought can be carried out, by analogy, parallel to the study of the above-mentioned unifying factors, as well as of themes regarding quantum mechanics and synergetics, such as: chemical bond, structure of matter, crystallography, spectrography, cooperative phenomena, dynamic instabilities, breaking of symmetry, and self-organization of dissipative structures in non-equilibrium. There is currently a tendency to include among the latter biological, economic and social systems: all the systems, that is, in which order may arise out of chaos.

Unifying factors

Finally, the purpose of this study falls naturally within the framework of the renewal of epistemology of the sciences, as promoted recently by Prigogine and his school:

> "Today the limits of what we can call Galileian idealization of nature appear within physics itself and the need for a description of reality that manages to integrate coherently two fundamental truths – every description of nature is produced by man and the man who produces it is produced by nature himself – is imposed by the internal development of science.
>
> The broadening of scientific theory, induced by current advances in physics, enables us to resume – on different bases – the eighteenth century program for a unified science which is both technology and philosophy of nature. This gives us reason to hope for a reduction in the conflict between the 'two cultures' that weighs so heavily on western thought today".
>
> (Prigogine and Stengers, 1979)

The "two cultures"

This study aims to outline criteria for creating a more coherent interfacing between natural sciences and the science of man. Still, suggesting analogies or proposing correlations among the phenomena characterizing these two cultural sectors implies the use of words and concepts which, transplanted from the scientific to the humanistic field, often end up being utilized, almost of necessity, in an improper way, or may even lose their original meaning. "Nomina nuda tenemus", declares Umberto Eco: we deal with bare terms. And we are jealous of this task, and careful that everyone performs it as accurately as possible, in his own particular area of specialization.

Thus it can be dangerous to pass the confines of the territory belonging to the application of scientific language and the results of the experimental method. This eighteenth-century-style attempt exposes me to criticism from colleagues and scholars in both the scientific world and the world of humanism: in my trespassing the former will be inclined to see inaccuracy and heedlessness, the latter presumptuousness and interference. Let me ask for mercy from the start: I'm running a risk, not imagining benefits.

This treatise is divided into six chapters.

The introduction which, as we've seen, established the purpose and defined the investigative methodology, is followed by a brief opening chapter. Without going into detail, this presents an overall picture of the problems under discussion. Entropy and information, symmetry and order are suggested here as unifying factors in understanding the relationship between man and natural structures: a relationship controlled by the instinct of self-preservation and enriched or plagued by contradictions that lead to ambiguity.

Ambiguity

The second chapter is dedicated to the natural structures. These structures are grouped according to quantum laws, modularly, beginning with atoms. They may be found near thermodynamic equilibrium or they may be removed from equilibrium by external stresses. In this chapter the quantum model of the atom is presented and the grouping mechanisms of atoms in molecules are analyzed. Formation of the covalent chemical bond between two atoms in a molecule, in particular, is visualized by utilizing ambiguous modular structures borrowed from the design. In dynamically perceiving the ambiguity inherent in these undecidable, unrealizable but not necessarily unreal structures, one grasps spontaneously both the resonant nature of the chemical bond and the concept of primordial information or unit of information – the *bit*. To make the chapter easier to read, certain technical details are described in Appendix I.

Symmetry and symmetry breaking

The third chapter, supplemented by Appendix II, deals with symmetry and the breaking of symmetry. Where these values – belonging to science, to perception and to art – converge, a conflict arises which, once again, expresses itself dynamically in ambiguity. The breaking of symmetry caused by the electromagnetic field applied in molecular spectroscopy experiments (the hydrogen or ammonia molecule, for example), and the dynamics that accompanies it, are useful reference terms in approaching an understanding of the mechanisms of the dynamic perception of ambiguous structures.

Entropy, information

The fourth chapter is dedicated to entropy and information. It includes an introduction to the first and second principles of thermodynamics, and to the connections between physical and informational entropy. These relationships seem useful in analyzing the structure of chemical-physical systems such as binary alloys. This analysis, conducted in Appendix III, is then extended to vastly more complex systems such as language – musical language and genetic language.

Dynamics of ambiguity

The fifth chapter deals with the problem of the dynamics of the perception of ambiguous structures and the formation of visual thinking, in the dynamic instabilities of dissipative structures removed from thermodynamic equilibrium. This chapter offers a phenomenological model of the mechanisms of the perceptive process, based on observations developed in the preceding chapters. Subsequently, a criterion is suggested for extending the definitions of symmetry and the breaking of symmetry used in physics, to poetry, the visual arts and music.

The main conclusions of the treatise are brought together in the sixth chapter. It is such a brief chapter that it would be difficult to propose a synthesis of it here.

The road to hell is paved with good intentions. In the hope of warding off any exhortations to go follow it, I have compiled, in extremis, a glossary. Intended primarily for readers with a literary, philosophical or artistic background, this fruit of good intentions could, in small doses, also prove to be useful to younger readers of technical extraction: at any rate, it should be generally harmless.

In concluding this premise, let me extend an invitation above all to those who, having already bought this book, are at this point thinking of postponing the reading of it sine die: before closing it and relegating it definitively to the shelf on the upper right, try reading at least a couple of the seven inserts on "Synergetics", on "Symmetry, Simplicity and Order", on "Symmetry, Information and Ambiguity in Quantum Physics and Design", on "The Musicality in Virgil", on "Exchange Symmetry", on "The Game of Broken Symmetries in the Industrial Trademark" and on "Selfsimilarity in Nature and Culture: The Fractals": these illustrate, simply and with the use of examples, those concepts developed in this premise and in the body of the chapters to which the inserts pertain.

These inserts were added to the manuscript last, but they should be read first.

Insert

Synergetics

> And everyone, rising on tiptoe, turned to look in the direction from which the unexpected arrival was heralded. All rising, they saw neither more nor less than if they'd kept the soles of their feet on the ground; but as it was, they all rose (...) this mass, having the greater force, can dispose of it at will, and thus each of the two active parts uses every means to attract it to their side, to possess it: they are practically two warring souls, battling to enter into that body and animate it. They compete to see who can spread the rumors most apt to stir up emotions, to direct movements favoring the cause of one or the other; to see who is better able, at the right moment, to find the news that may fan the flames of anger, or smother them, that may awaken hopes or terrors; who is able to find the cry which repeated by the majority and the strongest, will at the same time express, bear witness to and create the vote of the majority, for one side or the other.
>
> Alessandro Manzoni

Synergetics, or cooperation, is an interdisciplinary discipline which aims to unify culture. *Synergetics*

Fifteen years ago, Hermann Haken, in giving the go-ahead to a series of international conferences and publications on synergetics, coined this ancient neologism. Today, practically every physicist has put aside the diffidence with which, in a cultural environment, whatever is presented under the aegis of a new name is generally received.

Synergetics is concerned with behavior, with stability conditions, with dynamic evolution and with the nature of instabilities that may take place in complex systems removed from thermodynamic equilibrium, by external fluxes of energy and matter.

Prompted by these fluxes, the moduli composing such systems – atoms, molecules, cells, sensorial stimuli, animals and even human beings – interact among themselves in non-linear fashion. And they explore, try out, compare and weigh possible collective types of behavior of the structure, competing with each other. Until, as the external fluxes increase, the critical moment of choice is reached. At that point, only the slightest effort is needed to create a loss of balance, a fluctuation is enough, and the catastrophe happens: the system's symmetry is broken, and one well-defined collective type of behavior – mode – prevails dynamically over all the others. *Non-linearity*

A Darwinian-type mutation thus occurs, during which the structure organizes itself, stabilizing the collective mode that was chosen. Supported by the cooperation among the interacting structural moduli, this mode subjugates the structural moduli themselves and, growing steadily at the expense of the other modes, emerges victorious from the intermodal competition as a macroscopic order parameter, as the "organizer" typical of the structure's macroscopic behavior. (Haken, 1981).

Cooperation and competition

Order parameter

At the beginning of the fifth chapter we will propose a qualitative analysis, in synergetics terms, of the cooperative mechanisms that can lead to the onset of a cadenced rhythm – and thus of order – in the applause, disordered at first, rendered at the end of a fine concert: for these mechanisms to feed on themselves and become effective however, the system's control parameter, that is the audience's enthusiasm and the general wish for an "encore", must exceed a certain threshold.

Forms of order of a cooperative origin can also be found for example in the fluttering of a flag (here too however the wind speed, which assumes the role of a control parameter, must exceed a certain threshold), in the striated pattern sometimes formed by clouds in the sky or sand in the desert, in the sound of breaking glass, in the space and time periodicity of the concentration of chemical products formed in autocatalytic reactions, in the structure and dynamics of vortices in a river.

Control parameter

Not to mention the coordinated darting of a school of fish when a predator appears, the conformism of Manzoni's crowd, the formation of visual thinking or the establishing of political opinions; or in a slightly more distant perspective, the whims of fashion, the birthrate and the development of urban centers, and even the functional order underlying vital phenomena.

In short, the methods of synergetics, which are also the methods of non-linear physics of cooperative phenomena, enable us to interpret many phenomena occurring in systems which apparently have nothing in common with each other, using a limited number of mathematical models.

Cooperative phenomena

Thus, for example, the same system of non-linear differential equations describes both the dynamic instability which leads to the establishing of space-time coherence in an electromagnetic field in a laser cavity, and the dynamic instability which is a prelude to ordered convective motion in a fluid heated from below.

Generally speaking, the difficulties that must be overcome to formulate the mathematical model of a specific system away from equilibrium and to resolve the non-linear differential equations that schematize behavior and instability of the system itself, are considerable. It is to be hoped that the mathematical study of non-linearity, perhaps finalized to developing algorithms that are suitable and easy to use, will in the near future occupy a greater number of scholars than it has in the past.

Fortunately though, thought formation (visual thinking, for example) at the instant of perception is included among the dynamic instabilities that can be studied from the point of view of synergetics: consequently, what we experience in perception can offer, albeit at a sensorial level, a reasonably faithful understanding of the nature of processes which lead to the selforganization of ordered structures in the most diverse of natural systems, when they are removed from equilibrium.

Selforganization

Chapter One

Ambiguity in the Cultural Relationship Between Man and Natural Structures

πάντα ῥεῖ

Heraclitus, 6th century B.C.

The relationship between man and nature is characterized by the instinct of *conservation*, inherent in the actions and manifestations of all living species, and by the *transformations* which take place in all natural structures. *Everything flows*, Heraclitus wrote twenty-five centuries ago: and those two words remain, rooted beside each other, almost as if they were beating out, with continuous symmetry, the rhythm of time and of all transformations that time begets. The conflict between conservation and transformation is settled dynamically with that "panta rei", in an expression laden with *ambiguity*. Everything flows, in fact, but the fact that everything flows does not change. We realize this in contemplating the unvarying course of a river, indifferent to the passing of time, or the undertow, ever-changing in its regular, incessant rhythm.

Conservation
Transformations
Ambiguity

Knowingly or instinctively, we are all searching for something which remains the same while it changes.

But what is it that changes, yet remains unaltered?

Symmetry, a phycisist would reply, thinking of what does not vary in change, and in particular of the constants of motion of the structures near thermodynamic equilibrium on which he is accustomed to experiment. *Information* or *order* would be the answer of the biophysicist, who conceives of life as "a data-processing system – a structural hierarchy of functional unities – which through evolution has acquired the ability to store and process the information needed for reproducing itself with accuracy" (Gatlin, 1972).

Symmetry
Information

It is usually believed that symmetry and order are equivalent to each other as structural attributes. But this is wrong.

Order

For example, the term "order" is used by Arnheim in his essay on Entropy and Art (1974) as "an objective description of the simplest, most symmetrical, most regular form". The unfocused correlation thus suggested among order, simplicity and symmetry contributes, in my opinion, to creating the "babelic confusion" the author himself complains of when he exclaims that "someone or something has confused our language".

One may agree with Arnheim regarding the affinity between order and simplicity, but must differ on identifying order with symmetry (see Insert to Chapter One).

Symmetry as invariance

Symmetry may be defined in fact as *invariance under transformations*, and results from the impossibility of perceiving certain transformations, that is, from the "impossibility of measuring certain characteristic quantities of the natural structures" (Lee, 1968). Vice versa, order offers a *measurement of the observable correlations in the disposition, the sequences and the dynamics of the modules constituting the structures themselves.*

Symmetry versus order

Consequently, far from being identical with each other, symmetry and order are, if anything, in conflict. Landau and Lifshitz write (1967):

> "In most second order phase transitions, the more symmetric phase corresponds to the higher temperature and the less symmetric to the lower temperature. In particular, a second order transition from an ordered state to a disordered state generally takes place following a rise in temperature."

And likewise, Shubnikov and Koptsik (1974):

Symmetry and entropy

> " ... we also note that the behavior of symmetry as a function of the state of an isolated system is correlated to the behavior of its entropy. Highest symmetry ... is attained in the state of equilibrium of the system"

which – as we will see in the following chapter – is also the state in which entropy, meaning disorder, is greatest. Glansdorff moreover warns (1977):

> "The existence of [such] sources of order is not in contradiction with the second principle of thermodynamics, in spite of its being called the principle of energy degradation. Actually it's enough to remember that this qualification, associated with the property of increase of entropy, is justified only for isolated systems. This qualification doesn't extend to evolutions under constraints, either for closed systems or for open systems, that is for systems that can exchange matter with their surroundings."

It is no wonder therefore if Arnheim's originally semantic error becomes dangerously widespread when – in far off 1971 – the author, assimilating symmetry with order, ends in fact by changing the sign at the concept of order in the analysis, in thermodynamic terms, of the relationship between entropy (or disorder) and art.

Entropy is a characteristic function of the thermodynamic state of a system. As such, it also furnishes a measurement of the transformation that the system undergoes during its evolution.

In panta rei, at the meeting point between entropy (or transformation) and conservation, one perceives the coexistence of two aspects – incompatible with each other – of the same reality. One perceives, that is, ambiguity. But ambiguity also appears at the meeting point between symmetry or indiscernability on one hand, and information or the removal of uncertainty on the other: in this way it ends by permeating the behavior of natural structures and man's cultural relationship with them.

Entropy and *conservation*, *information* and *order*, *symmetry* and *ambiguity*: concepts are associated to these key words which take on a unifying role in interpreting the complex relationship between man and natural structures.

The proposed connections among these unifying factors may be seen synoptically in the scheme of Fig. 1. Here the natural structures are classified in two groups:

Near equilibrium structures

1) **Structures near thermodynamic equilibrium.** The spontaneous evolution of these structures is associated to disorganization, disorder, incoherence, uncertainty and entropy, which increase gradually until they reach, compatibly with the bonds, a maximum value. The state realized in this way corresponds to a static equilibrium at the macroscopic level.

A classic example of such structures is a gas made up of molecules closed in an insulated container, composed of two equal and communicating compartments. If the gas molecules are all collected initially in the compartment on the left, after a certain time they can be found half on the left and half on the right. Once they have spontaneously achieved this arrangement, less ordered than the original one, the system remains in equilibrium: its entropy has reached a maximum value.

Entropy

Dissipative structures

2) **Dissipative structures** (including biological structures) removed from thermodynamic equilibrium by external flows of energy, information and matter. In these, "specific kinetic laws permit the construction and maintenance of a functional and structural order" (Glansdorff and Prigogine, 1971); "a critical distance from equilibrium, that is, a minimum level of dissipation, is needed to maintain them" (Prigogine, 1979).

Order

The Rayleigh-Bénard instability

A recurring example of dissipative structure is that of a thin layer of liquid closed between two plates in such a way that a free airgap does not form (Rayleigh-Bénard capsule), and heated slowly from below. For small temperature gradients, that is, in the immediate neighborhood of thermodynamic equilibrium, heat diffusion is sufficient to dissipate the thermal energy supplied, without collective motion of the liquid (which macroscopically remains in a state of rest) intervening. But what happens when the temperature gradient that has been set increases, and with it, the flow of energy into the liquid? To a degree determined by the magnitude of this gradient – which acts as a *control parameter* of the dynamic behavior of the liquid – the system, still below the boiling point, moves away from thermodynamic equilibrium. It becomes increasingly difficult for heat diffusion, by itself, to guarantee the dissipation of the thermal energy supplied. Until, unexpectedly, another channel opens, in *competition* with the diffusive channel, through which the energy supplied may be dissipated: and abruptly, at the Rayleigh-Bénard instability, a macroscopic collective motion is triggered. The hottest parts of the liquid, those closest to the heat source, expand, and because of Archimede's buoyancy tend to float, moving upward. Here they are cooled down, thus falling again toward the bottom. In this way, a process of heat transport by convection is established, depending on a collective motion of the liquid, and characterized by cooperation at the molecular level. This collective movement is produced thanks to the kinetic energy developed by the work of the forces caused by the warmer mass's tendency to float, in the reverse gradient generated by the flame.

Control parameter

Competition

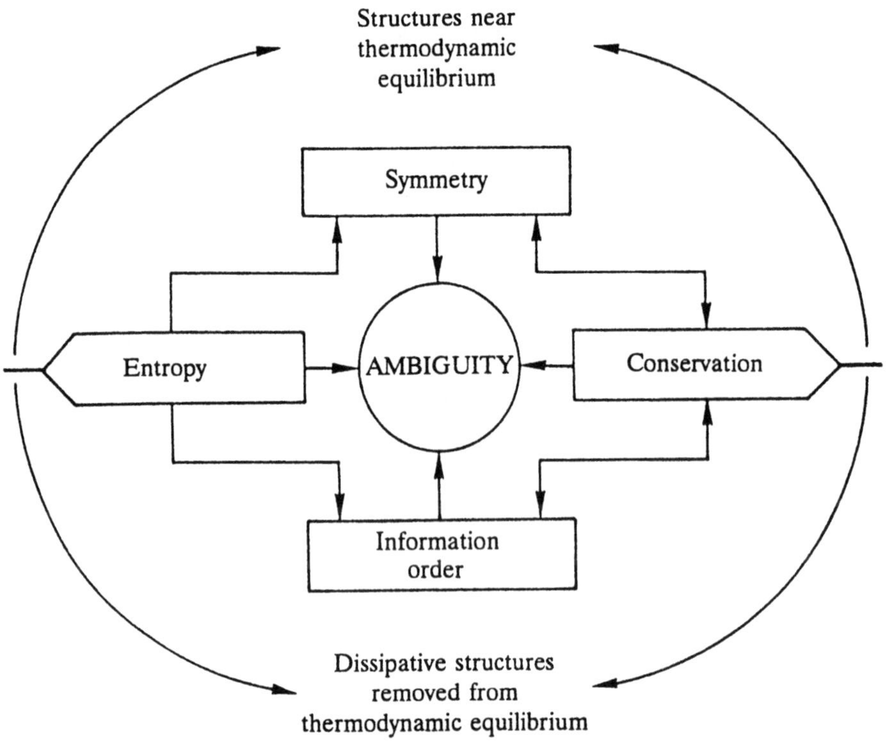

Fig. 1. Entropy and conservation; symmetry and order, information and ambiguity: these are the factors which assume a unifying role in the cultural relationship between man and natural structures. In the entropic evolution of the structures near thermodynamic equilibrium, what is preserved is symmetry. Vice versa, symmetry singles out quantities that are preserved (constants of motion). In the neg-entropic evolution of the dissipative structures, far removed from thermodynamic equilibrium, it is information and order that remain. Vice versa, information fosters a correlation among the structural modules – order – which tends to be perpetuated (think of the instinct of self-preservation in individuals and biological species, for example). The conflict between entropy and conservation, between symmetry and information is resolved, dynamically, in ambiguity

Fig. 2. To start off a convective flow the forces must be unbalanced. Let us take an experiment in which a thin layer of fluid between two flat surfaces is heated from below producing a gradient of temperature and density. If some of the hot fluid in the lower section of the layer is transferred toward the upper part, it enters a region of greater average density and is therefore subjected to buoyancy force towards the top. In the same way, if some of the cold fluid near the upper part of the layer moves downwards, it becomes heavier than the surrounding environment and tends to descend even further towards the bottom. The buoyancy force is contrasted by viscous resistance and by heat diffusion, which tends to bring the temperature of the transferred portion to the same level as that of its surrounding environment. The relative importance of these contrasting forces is measured by the Rayleigh number. Convection begins when upward buoyancy exceeds the dissipative effects of viscous resistence and thermal diffusion, that is, when the Rayleigh number goes beyond a critical level. (From M. G. Velarde, C. Normand: Le Scienze, no. 145, September 1980, p. 70)

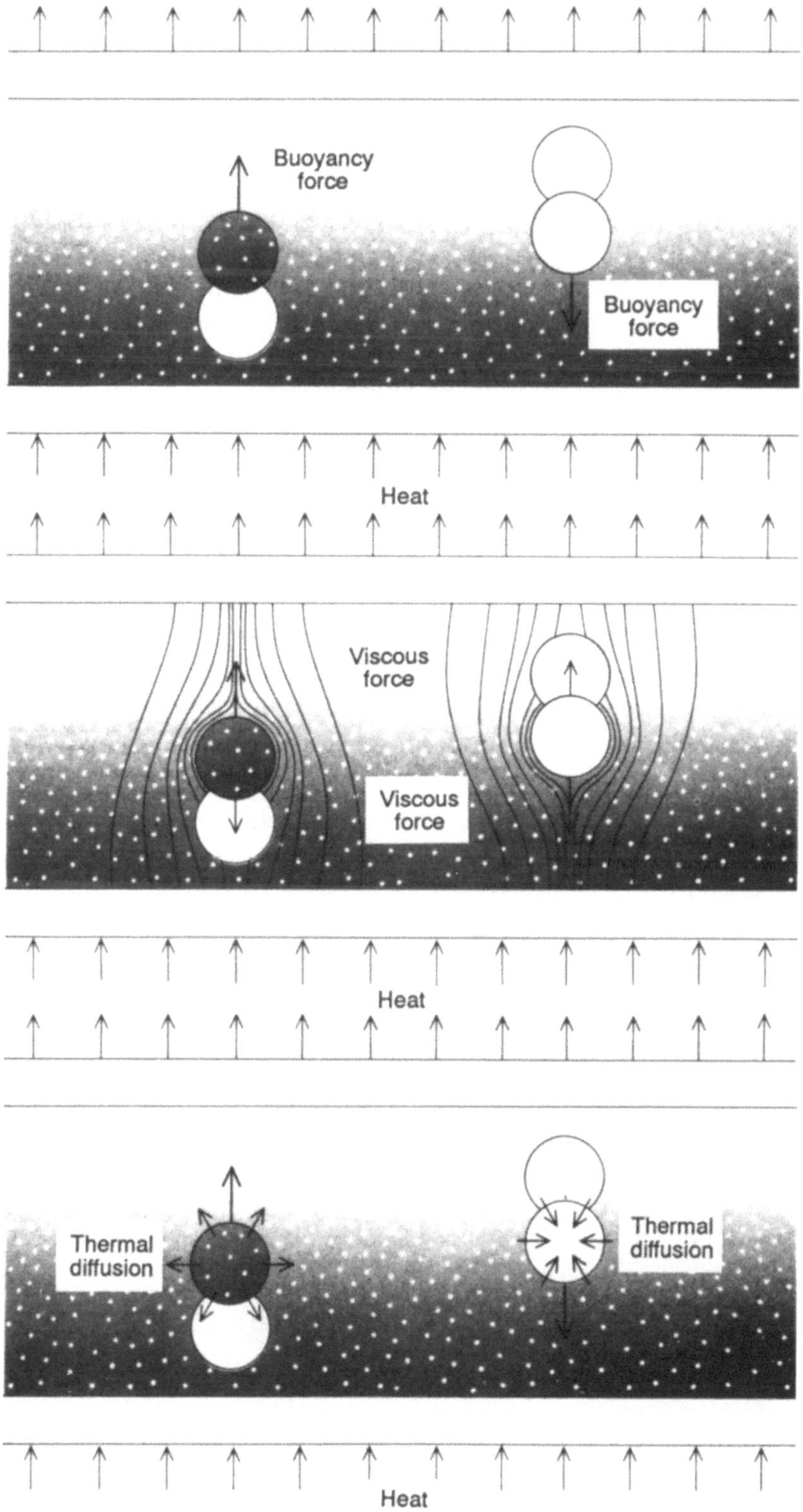
Buoyancy force
Buoyancy force
Heat
Viscous force
Viscous force
Heat
Thermal diffusion
Thermal diffusion
Heat

Fig. 2

The movement is obstructed both by heat diffusion, which tends to level the temperature field and thus to cancel out the density gradient responsible for the buoyancy, and by the viscosity of the liquid. On the near side of instability, the energy supplied to the structure deteriorates into disordered thermal agitation energy: fluctuations formed by correlated micromovements of small groups of particles of the liquid, warning signs of convective movements, occasionally arise, only to recede immediately. Beyond instability, instead of receding, these fluctuations – driven by that source of order which is identified with the deviation from the thermodynamic equilibrium – amplify (Fig. 2) and they become stabilized in an *ordered structure, dynamically ordered as well* (Fig. 3). This new structure organizes itself at the expense of externally supplied thermal energy: part of this energy is transformed into kinetic energy of ordered macroscopic fluid strings; the formation of these strings breaks the translational symmetry typical of the homogeneity of the liquid's structure; the magnitude of the velocity and temperature field associated to the fluid strings is the *order parameter*; the persistence of the fluid strings, improbable or even impossible in an isolated system, is subordinated, as mentioned above, to the availability of a thermal power that the liquid's viscous structure dissipates to maintain the velocity and temperature gradients that are established among the fluid strings themselves: once the burner has been turned off, the availability of thermal power disappears, and with it the ordered and cooperative convective movement; until the temperature has levelled off only the incoherent mechanism of heat conduction by diffusion remains.

Order parameter

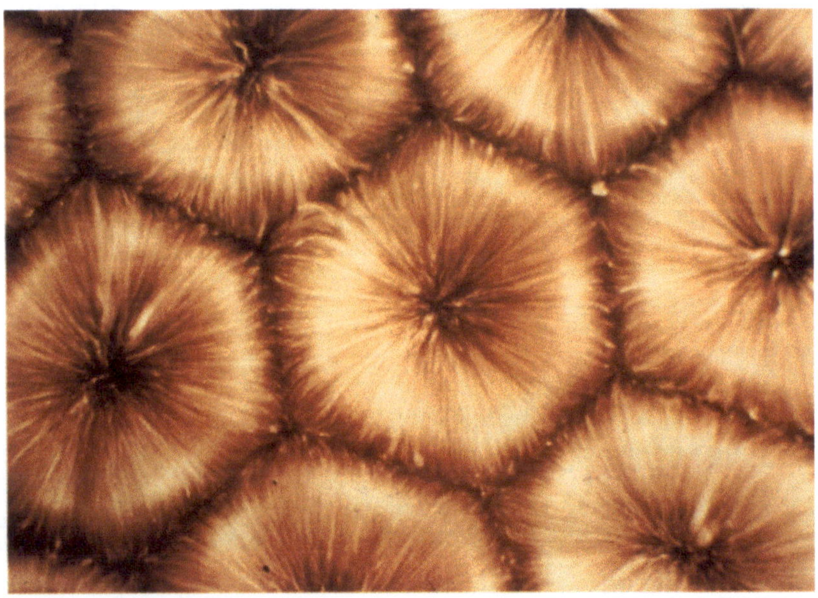

Fig. 3. Surface-tension-driven cells of convection beyond the Bénard instability in silicon oil. Light is reflected from aluminium flakes suspended in the oil. The oil flows upwards at the center of the hexagonal cells and moves towards their periphery, where it flows downwards. (Courtesy of M. Velarde)

The self-organization of the convective movements of a fluid, controlled by the inverse thermal gradient in the Rayleigh-Bénard capsule, has characteristics analogous to that of the upward movement of smoke and hot air in a chimney hood.

We could also cite the distribution of fog, which tends to thicken and condense predominantly along the peripheral ring around large urban centers. Or the candles lit on a table laid with care, which can help to cut down the smoke and create a refined atmosphere.

Now let us return to the synoptic scheme of Fig. 1.

Entropic evolution, equilibrium and symmetry

What is preserved in the transformation or entropic evolution of structures near thermodynamic equilibrium is symmetry. Vice versa, symmetry singles out observable quantities which are preserved.

Neg-entropic evolution, nonequilibrium and order

What is preserved in the transformation or neg-entropic evolution of non-isolated structures away from thermodynamic equilibrium, including living systems, are information and/or order. Vice versa, in these structures order tends to perpetuate itself, that is, to be preserved.

To anticipate a subject that comes up often in this treatise, it seems important at this point to emphasize that ambiguity plays a central role in resolving dynamically the contradictions inherent in the cultural relationship between man and the object of artistic "consumption" or even of scientific observation.

Ambiguity

Ambiguity could in fact be defined in terms of *coexistence, at a critical point, of two aspects or schemes of reality which are mutually exclusive* and which have become physically observable – in the perceptive process or during a spectroscopic observation, for example – by means of a transition related at times to a characteristic frequency (Plate II).

Actually, it seems possible to propound important analogies between the behavior of dissipative structures away from thermodynamic equilibrium and the process of perceiving an ambiguous figure. (To avoid misunderstandings, we are referring here and hereafter to a figure that is understood to be interiorized, after the preanalyses made by the retina and the optical ganglions).

Gestalt structures

Let us proceed in orderly fashion. Take the gestalt structures in Plate III, and observe one of the two blocks with particular attention.

The sensorial stimuli this figure gives rise to are controlled by the observer. He endeavors to organize these stimuli, to keep them from slipping his mind, arranging them in correlated, coherent schemes. In the early phase of the perceptive process, control of the stimuli, still scanty, permits correlation only locally and in a temporally discontinuous way of a small number of structural modules scattered here and there: the latter bring to mind, in incoherent and practically disordered form, the explorative stimuli "processed" by the observer. And the structure appears to be two-dimensional and center-symmetric. What happens however if control of these stimuli exceeds a critical threshold, thus enabling the observer to tune himself into the field of attention produced by the signs which define the structural modules, and by the dynamic components associated to them?

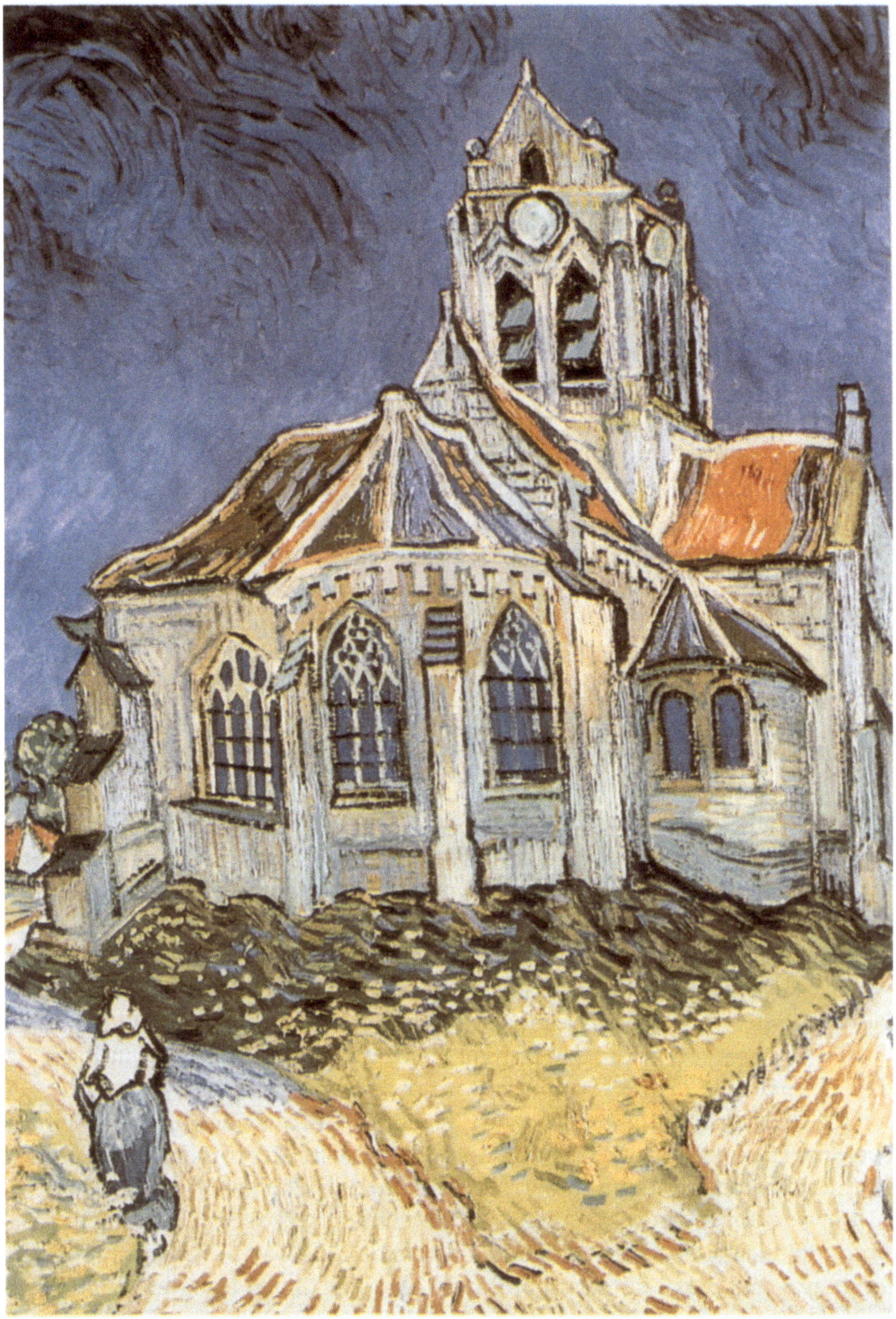

Plate II. L'église d'Auvers, V. van Gogh (1890), Musée de l'impressionisme, Paris

This is one of Van Gogh's last works. With his incisive stroke, the artist is able to grasp the essence of structures and bring out their rhythm. What is the secret of the movement that animates his painting? Many architectural elements of l'église d'Auvers, if considered individually, can be *read* or "visually thought", each in two ways which are mutually exclusive and linked by a dynamic transition which consists of a periodic alternation of inversions of perspective. For example, if you look at the wall on the extreme left long enough, it seems to shift, so to speak, *zagzigging* (instead of *zigzagging*) downward, so that the second and the fourth horizontal edges appear to be *on the near side* (rather than *on the far side*) of the ramps they define. More or less the same thing happens with the steps in Schroeder's staircase (Fig. 4); it is as if each of the two edges may be perceived in two positions that are symmetrical with respect to the plane of the picture. Similar sensations are produced when perceiving other structural elements, for example in relation to the window-sills as well as the sides of the four-light and three-light windows, and even the furrows (or protuberances) of earth towards the church. All this gives the work a variety of hidden symmetries, which the observer feels the urge to break, triggering periodic transitions each having its own characteristic frequency: this frequency – it is important to point out – depends, ceteris paribus, on the size of the painting's structural modules, or better, on the visual angle subtended by the painting (Borsellino, Carlini, De Marco, Penengo, Riani, Trabucco, Tuccio, 1982). This is one of the objective reasons for which the aesthetic sensations stimulated by the original work are usually far more violent than those offered by a smaller-sized reproduction of it (Paris vaut bien une Messe!). This is particularly true of Van Gogh, whose plastic stroke helps heighten the sense of relief. To perceive all the changeable combinations pertaining to the periodic transitions that define the characteristic frequencies of this structure, full of ambiguity, would require an observation time even longer perhaps than that needed by Van Gogh to transfer them instinctively into the painting. Optical illusions of this kind (but are they really illusions? Isn't it perhaps true that "verum et factum convertuntur"?) are not limited to ambiguous, bidimensional figures. Surprisingly, they may even be suggested by what Montale would call "the most static of the arts (architecture)". One needs only observe, in the cloister next to the church of San Carlino at the Quattro Fontane (Borromini, mid-17th century) in Rome, the connective elements between flat and convex surfaces above the coupled columns of the lower order. Or note, first in the frieze and then down along the sides of the lateral pillars in the apse of another masterpiece of baroque Rome, St. Ignazio, the silent play of hollows and

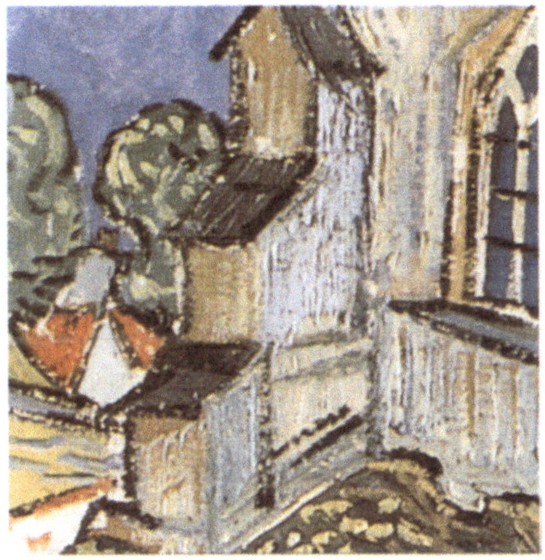

projections, which spring out majestically and exchange their roles. Throughout this treatise, the perception of an interiorized ambiguous structure is interpreted in terms of a dynamic instability of the sensorial stimuli, controlled by the observer

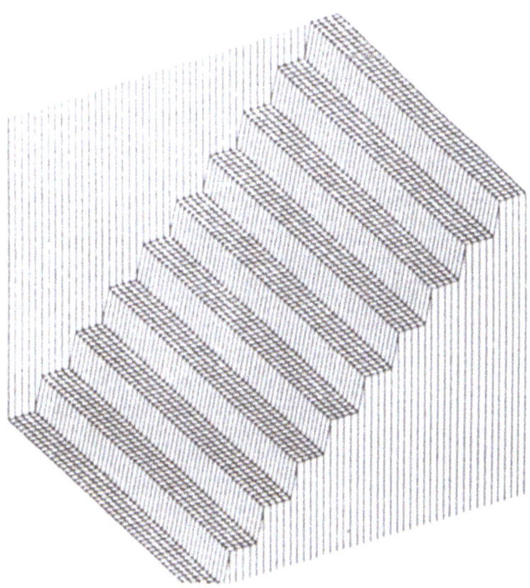

Fig. 4. Schroeder's staircase (from Melvin L. Prueitt: *Computer Graphics*. Dover Publ. Inc., New York 1975). After a few seconds of observation, this staircase seems to turn around; the mind tries to establish a pattern, but is left undecided when the shifted pattern is presented to it

The symmetry of the structure having been broken, the stimuli, which were formerly scattered stochastically, as if by magic arrange themselves dynamically, depicting the visual thinking, and are re-irradiated collectively by the figure. That is, they begin to oscillate en bloc, with a frequency which is on the average close to the characteristic frequency that the artist or designer wished to give the structure by imposing specific morphologic interactions among the structural modules composing the structure itself.

Perception and physics

As mentioned in the introduction, one reason which justifies dedicating entire sections of a book to the analysis of these undecidable and non-concretizable structures stems from the fact that the phenomenological model we are about to elaborate for the mechanism of their dynamical perception apparently involves, at the level of analogy at least, a certain number of chapters fundamental to physics – symmetry and the principles of conservation, broken symmetries and the structure of matter, spectroscopic observation, thermodynamics and entropy, the theory of information and irreversible thermodynamics, cooperative phenomena and self-organization of structures in non-equilibrium: these subjects will be confronted in the following chapters. Actually, one could say that to grasp the essence of these problems at the physiological or intuitive level, those few seconds needed to perceive and to think visually of an ambiguous structure may suffice.

In conclusion, during the perception which introduces formation of the idea and development of thought, and especially during the process of perceiving the simplest of the "forms-non forms" (Thom, 1972) belonging to the class of ambiguous structures, all the unifying factors that we have mentioned up to this point are present, combining to produce our own personal image of reality. In analyzing the perceptive mechanisms for these structures – leaving the neurophysiological aspects apart – we will limit ourselves to bringing out the *qualitative* functional correlations among the above-mentioned unifying factors. This analysis constitutes a dominant theme, forming the basis for the six chapters in which the treatise is divided.

Insert

Symmetry, Simplicity, Order

Once symmetry is broken by control, the idea appears. And then order follows. And thought, which governs all, in which the idea, in its simplicity, unfolds. The idea of *simple* is in the turn it takes.

Sem-plice or folded once

Under the entry "semplice" (simple) in the Zingarelli dictionary we find: "combination of *sem-*, once, and of a derivative of *plectere*, to fold"; also: "consisting of a single element".

Concentrated in the etymology of words we often find, to a degree that is always surprising, the ancient wisdom of individuals who, living in immediate contact with nature, knew instinctively how to comprehend its laws and behavior.

Many systems exist in nature which, during their evolution, *take a certain turn*. We refer to chemical-physical, biological, economic and social systems which, having reached a critical point, undergo abrupt structural modifications or dynamic instabilities, real revolutions, which mark radical *simplifications* in their way of being or their behavior.

Take a magnet, for example. A magnet, we know, attracts iron. The characteristic of attracting iron is linked to a measurable, macroscopic physical quantity: magnetization. Magnetization is the result of a multitude of microscopic addenda, as many as there are magnetic atoms (iron or nickel atoms, for example) within the magnetic material of which the magnet is made. Each magnetic atom behaves like the needle – greatly reduced – of a compass. But how does the overall magnetization depend on the orientation of the individual needles?

Let us refer here to two extreme situations (Fig. 5 a and c).

If, inside the magnet, all the needles point *in orderly fashion* in the same direction, their contribution to magnetization is summed up coherently, the magnetization takes on a maximum value and the magnet attracts the iron (Fig. 5 c).

If vice versa the needles point *in disorderly fashion* each in a different direction, their contribution to the magnetization is incoherently summed up to zero, the value of magnetization is null, and the "magnet" is a magnet no longer, since it no longer attracts iron (Fig. 5 a).

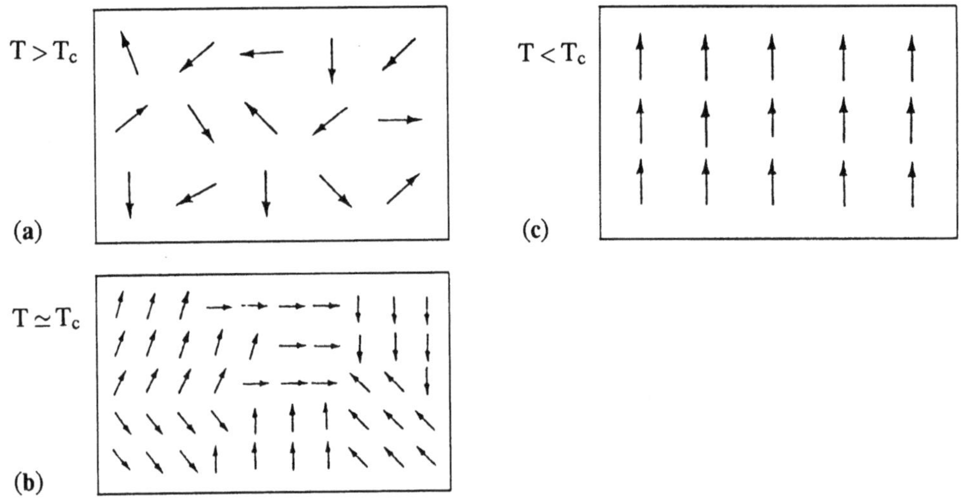

Fig. 5 a–c. In a magnetic material the configuration of the magnetic, atomic needles is disordered above the critical temperature T_c, (**a**), and is ordered below the critical temperature T_c, (**c**). When, starting from $T > T_c$, the temperature is lowered to values just above the critical temperature, $T \sim T_c$, (**b**), fluctuations in the orientation of the needles, instead of diminishing, are correlated and tend to involve wider and wider spatial areas. If the temperature is reduced further, $T < T_c$, (**c**), one of these fluctuations, growing quite large, pilots the magnetic needles contained in a spatial region of macroscopic size and, orienting them parallel to each other, assumes the role of parameter of order for the new ferromagnetic structure.

Ferromagnetic order and paramagnetic disorder

These two types of behavior occur, for the same material, respectively below and above a critical temperature, which indicates the transition point between the ordered phase (ferromagnetic) and the disordered phase (paramagnetic).

Above the critical temperature ($T > T_c$) the atoms of the material have a relatively high energy and toss the needles incorporated in them around, jumbling their possible orientation incessantly: describing the microscopic configuration of the paramagnetic phase in detail, in which each needle, deafened by the thermal noise, orients itself independently of the others, would be as arduous as it is useless.

Magnetization, the order parameter

Vice versa, below the critical temperature ($T < T_c$) the thermal noise, so to speak, isn't so high as to prevent the needles from communicating among themselves and from adopting an orientation in common: it's enough that a fluctuation appear, as if by accident, causing a group of nearby needles to orient themselves lined up in the same direction, for the behaviour of the magnetization to take, of necessity, *a certain turn*: all the other needles, in the field induced by the common orientation of the first group, will imitate them. And as the number of imitating needles increases, the field favoring the orientation of the remaining needles grows: until, in a macroscopic domain of the material all the needles arrange themselves in the same direction, and *a single ordering element*, magnetization, suffices to describe the magnetic behavior of the entire structure. Alea iacta est: the crisis is over, the choice is made.

The situation is similar to that which is created when in the evening, after disordered and aimless wandering around, a group of youngsters begins to feel hungry but disoriented in choosing one among the many "pizzerias" downtown: it's enough that two or three of them, understanding each other at a glance, make a choice and start off together, and the rest will fall in with them: as one man, everyone heads to Gennaro's!

Temperature, the control parameter

Thus *a parameter of control*, temperature, and a *parameter of order*, magnetization, play a fundamental role in the magnet's behavior. At the critical temperature T_C, the trend of the magnetization with the temperature takes a certain turn (Fig. 5 b): the situation of indescribable paramagnetic disorder, pertaining to zero magnetization, is substituted, as the temperature decreases, by a situation that may be described *simply*, by *a single element*, which is magnetization itself.

Dynamic instabilities

The structural modification just described is a second order phase transition. The latter is characterized by the fact that, at the critical temperature, the states of the two phases coincide ambiguously (unlike what occurs, for example, when water reaches the boiling point: at this point, the fluid that is undergoing a phase transformation of the first order is at equilibrium in two different states of aggregation – liquid and vapor). To whoever asks whether the phase realized at the critical temperature is ordered or disordered, one would have to answer: both. Tertium datur?

This transformation is in some ways analogous to the dynamic instabilities promoted synergetically by fluxes of energy, of momentum, of information and/or matter in the dissipative structures. These fluxes, which remove such structures from thermodynamic equilibrium, assume the role of *control parameter* of the dynamic instability; and they produce an effect similar to that brought on by a lowering of the temperature in para-ferromagnetic transformation, or by appetite in the group of youngsters: that is, acting on a specific fluctuation inside the structure, they control the possibility that the fluctuation may either regress or – *order parameter* – that it grow catastrophically large, reorganizing within the structure a new phase that is ordered, or correlated, even at a microscopic level.

Ordered macroscopic structures – whether they be dissipative, like those according to which the convective movements of a liquid in the Rayleigh-Bénard capsule are organized, or non-dissipative, such as a ferromagnet or a metallic alloy below the critical temperature – are, consequently, simple at a microscopic level as well.

Symmetrical macroscopic structures could appear at first glance to be simple and ordered: they may in fact give rise to the impression that only a few elements are needed to characterize them. Actually, these structures, if they were at most ideally symmetrical, would be so undifferentiated and immensurable that it would be impossible even to describe them; an ideally symmetrical structure is not simple, it has undergone no turn.

Idea and thought

Incidentally, thought itself, through which our relationship with natural structures is established and the revolutions which take place in them are experienced, rationally or emotionally, can be organized in an ordered and coherent scheme starting with a great number of disordered and incoherent sensorial stimuli when, gradually or abruptly, the control parameter of the stimuli exceeds a critical value. The dynamics of thought – *as an order parameter characteristic of the field of attention* – would thus consist in selecting, beginning with the disordered group of stimuli that are important a priori, the dominant stimulus which becomes emotion or idea, and thus, in its turn, thought.

The dynamics of thought

In this perspective, what occurs in para-ferromagnetic revolution (or "election") does not seem so different from what takes place when we must develop a theme or draft a report. Only after intense hidden activity, aimed at checking and correlating the many elements important a priori to the presentation, does the crucial element – the idea – insinuate itself, and the only truly representative order parameter – thought – emerge.

In brief, the idea may be explained simply: the idea of *simple* lies in the turn it takes.

Chapter Two

Natural Structures

At the point where nature stops producing its species, there man begins with the things of nature, with studying this nature, to create infinite species.

Leonardo

In the preceding chapter, it was suggested that natural structures be classified in two groups: structures near thermodynamic equilibrium, and dissipative structures.

All structures, whether they be natural or artificial, isolated or exposed to fluxes of energy, matter or information, have one characteristic in common: modularity. As Democritus suggested in the 5th century B. C., their construction is based on atoms, which are themselves grouped modularly to form the molecules, liquids, crystals, amorphous materials, or organic substances which compose living material.

Quantum laws

On the atomic scale, the atomic structures are regulated by quantum laws. The macroscopic properties of such structures and their behavior when acted upon by external fields may be interpreted thermodynamically and synergetically. Unfortunately these subjects are usually presented in esoteric terms or, as Brillouin would say, "in physicists' language, crammed with a jargon that only the specialists can understand": with consequences that range from incomprehension to indifference, from the short-sighted polemics of the two cultures to unjustified attitudes of diffidence and contempt.

We will try to avoid these pitfalls.

Atoms

The atomic elements will be presented first of all, to be followed by the simplest of the systems in which atoms are modularly grouped.

The molecule and the bit

We will dwell particularly on the nature of the mechanisms which govern the formation of the chemical bond that acts among the atoms in the molecules, on the "structure" of the informational bit and on the dynamics of the conflict between symmetry and order: in spite of appearances, these topics are correlated among themselves. Finally, assuming that during the process of perception the sensorial stimuli – once they have passed a threshold of instability – may organize themselves synergetically to form visual thinking, we will lay the bases for a possible analogy between spectroscopic observation of a binary molecular structure and the perception of an ambiguous structure.

1. Atomic Modules Constituting Molecular and Crystalline Structures

At the beginning of the 19th century, Avogadro formulated his famous hypothesis to put into logical perspective the laws established by Proust, Dalton and Gay-Lussac regarding chemical reactions, thus concluding once and for all a thousand-year-old ideological dispute on the nature – homogeneous or atomistic – of matter, while at the same time raising a concrete and complex problem for the generations to come: the problem of the structure of molecules and crystals, and that of the elements composing these natural systems, the atoms.

The atom

This paragraph presents the atom schematically: it is an object so small as to actually set a limit – etymological as well – beyond which divisibility is no longer possible, and at the same time, an element incredibly imbued with uncertainty, and therefore with symmetry.

According to Rutherford's experimental findings, the atom is composed of a nucleus and electrons. The nucleus, formed in its turn of protons and neutrons, is relatively heavy and astronomically dense. It is set in the center of the atom and has a positive charge which is a multiple of the elementary charge associated to a single proton. Attracted by the nucleus and surrounding it are the electrons; they are negatively charged: the number of electrons is equal to that of the protons in the nucleus, so that the atom is electrically neutral. Until the mid-1920s, the atom was envisioned as a planetary system in miniature. The size of the atom, which actually occupies a volume on the order of 10^{-23} cm^3, and that (vastly reduced) of the central nucleus, is however too small compared to the size of the objects we know for us to presume it possible to transfer sic et simpliciter the classical laws governing macroscopic phenomena to the world of atoms: in other words, it should not be surprising that the experiments regarding interaction between radiation and matter, carried out at the end of the last century and the beginning of the 1900s in the field of atomic spectroscopy and radiation, confirmed the inapplicability of the classic laws in the microcosm. In fact the validity of the classic laws had been confirmed only in the macroscopic field.

Spectroscopy

In order to interpret the experiments in atomic spectroscopy, it thus became necessary to substitute the new laws of quantum mechanics for the classic laws of mechanics and electromagnetism. The new laws lend themselves to easy application at the atomic level and, as might be expected, they tend to identify with the classic laws when the space-time sizes of the system considered assume macroscopic values.

Quantum mechanics

To appreciate the content of the quantum laws, it may be helpful to make a parallel analysis of the quantum behavior of the electrons in the atoms and the classic behavior of an artificial satellite in the earth's gravitational potential.

This analysis is carried out in Appendix I; the following summarizes the conclusions.

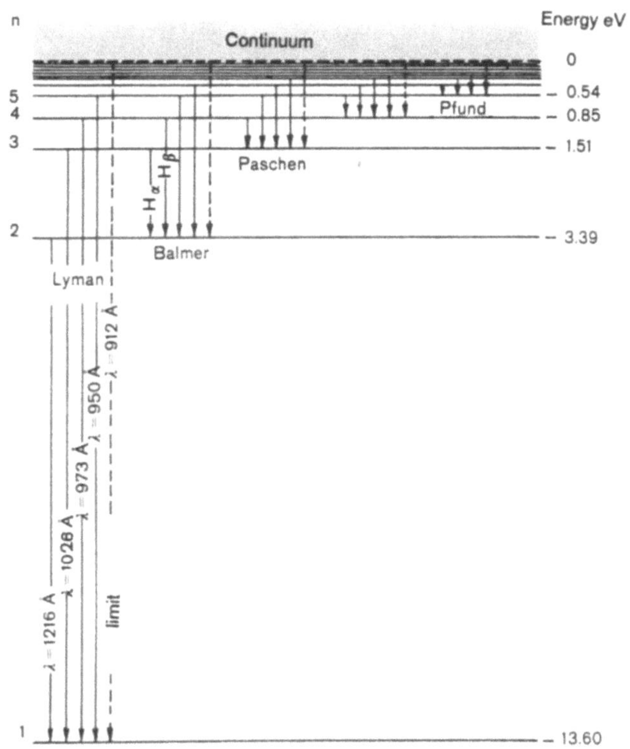

Fig. 6. The energy levels of the hydrogen atom are ordered in a discrete succession rather than according to a continuous distribution. Beginning with a minimum value (equal to -13.6 eV, for the ground state indicated by the principal quantum number $n = 1$), the possible energies of the electron bound by Coulomb forces to the proton grow and become equal to -3.39 eV for the first excited level (pertaining to the principal quantum number $n = 2$), -1.51 eV for $n = 3$, etc., until they vanish for the electron orbiting at enormous distances from the proton. In the hydrogen atom just as in any quantum structure, during transitions from a higher energy level E_m to a lower energy level E_n, an electromagnetic radiation with the frequency $\nu_{mn} = (E_m - E_n)/h$ – where h is the Planck's constant – is emitted. This is Bohr's equation. The wavelengths λ of this radiation are linked to the frequency by the relation $\lambda = c/\nu$, where $c = 3 \cdot 10^{10}$ cm s^{-1} is the speed of light. If the atom is placed in an electric field, the energy levels, except for the ground level, split up into so-called multiplets (Stark effect, Fig. 17). (From G. Caglioti: Introduzione alla fisica dei materiali. Zanichelli, Bologna 1974)

The hydrogen atom

Let us refer to the hydrogen atom, which is the simplest atomic system.

The only electron in this atom, attracted by the proton, has an energy which – unlike what occurs for the artificial satellite – may be fixed not as one pleases, but within a discrete succession of values that are characteristic of the atom (Fig. 6).

Energy levels

These possible values of energy (which arise as eigenvalues of Schroedinger's equation) are called *energy levels*.

State function

Once the particular energy level in which the electron is located has been established, the electron's existence around the proton is substantially independent of time and is defined not as an orbit, but in terms of *wave func-*

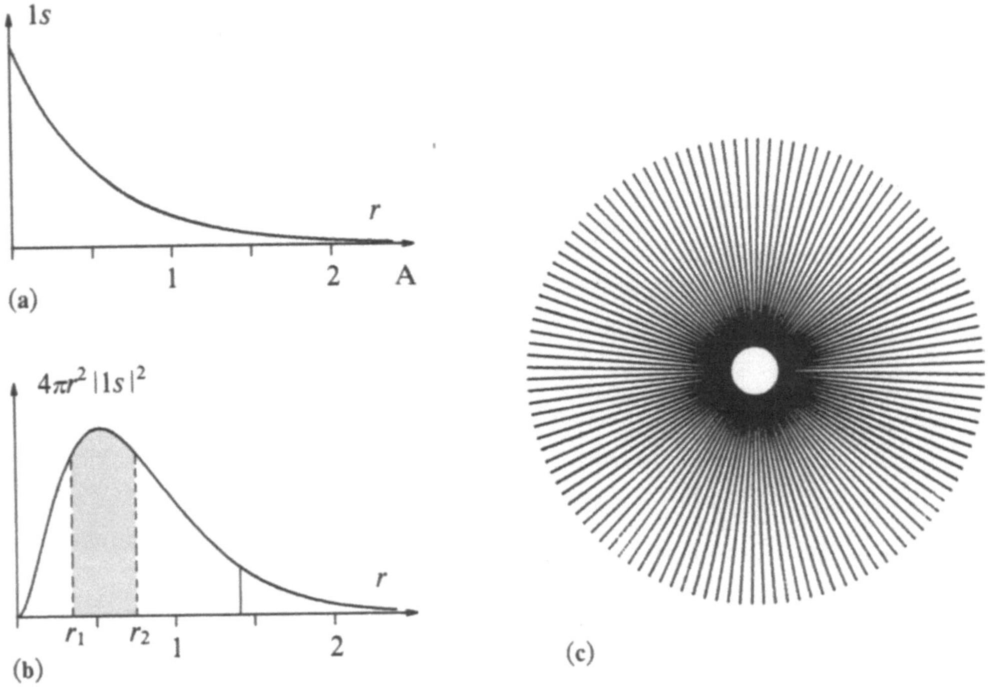

Fig. 7. (**a**) The wave-function $1s$, relative to the ground state of the hydrogen atom. (**b**) The electron's radial density of probability, $P(r) \equiv 4\pi r^2 |1s|^2$, at distance r from the nucleus. The shaded area represents the probability that measurement of the electron's position in energy's eigenstate $1s$ will give a value between r_1 and r_2. Note that this probablity is greatest near the hydrogen atom's Bohr radius, $a_1 = 0.53$ A. The total area defined by the curve and by the axis of the abscissas is obviously unitary. (**c**) Pictorial representation of the atomic orbital associated to hydrogen state $1s$. This orbital is defined as a space region – bound by a surface along which the square of the wave function is constant – inside which the probability of finding the electron is equal to 0.9. In eigenstate $1s$, the angular moment of the electron is zero: the electron hurls itself against the nucleus (represented by the white sphere in the center), without ever remaining trapped

tion or *state function* (or stationary eigenfunction of Schroedinger's equation). The state function associated to the aforementioned energy level allows us to evaluate the probability that the electron it describes be found, during a measurement, in any region of the space surrounding the proton.

Atomic orbital

While furnishing the greatest amount of information theoretically obtainable on the system, the state function permits the behavior of the electron near the proton to be described in probabilistic terms only. For example, the probability that the electron of the hydrogen atom in its fundamental state 1*s*, corresponding to the lowest energy level, be found in a sphere having a radius of $1.5 \cdot 10^{-8}$ cm centered on the nucleus, is 90%. This spheric volume, (Fig. 7), bound by a surface on which the state function is constant, is called *atomic orbital* associated to the ground state, since one is reasonably sure of finding the electron inside it.

Any further need for certainty regarding the location of the electron described by a stationary state could appear justified and at first glance gratifiable if the atomic electron behaved like a classical particle; actually, to fulfill this need it would be necessary, even within the classical limit, to set the starting conditions with great precision: this would hold back, in an ideally definitive way, the gradual spread of the inevitable initial uncertainty, caused by the deterministic evolution of the system. In fact, this need, if it refers to the electron in the most appropriate context of quantum mechanics, becomes definitely excessive.

Uncertainty

2. Binary Structures: Chemical Bond of the Electrons in Molecules

Natural systems are composed of groups of structural modules correlated among themselves.

In order to gain an understanding of them, it is desirable to identify some simple model-system which is open to a physical analysis that is as rigorous as possible, but at the same time representative of situations that interest the field of perception and art as well.

It seems to us that binary structures can satisfy these requirements.

Molecules and ambiguous figures

On the microscopic scale, we are referring to systems such as the electron in the hydrogen molecular-ion, H_2^+, or the nitrogen in the ammonia molecule, NH_3, as well as the spectroscopic observations that it is possible to carry out on these molecules. Concurrently, on the macroscopic scale, we refer to the "graphic condensation" of ambiguous figures, for example (see Plate IV, created expressly by Franco Grignani), and to the relationship we establish with these figures when they are perceived.

The chemical bond

This paragraph introduces the concept of chemical bond, with reference to the simplest of the binary structures, the hydrogen molecular-ion. The following paragraph deals with the problem of spectroscopic observation of binary structures – such as H_2 and NH_3 – by means of a resonant electromagnetic field capable of breaking the symmetry of the stationary states associated to these systems. The discussion of the process of spectroscopic analysis of these binary structures leads naturally to the concept of uncertainty removal or information: a concept which is presented in preliminary form in paragraph 4, and then developed further, together with the problem of the perception of ambiguous structures, in subsequent chapters.

What happens, then, when two elementary structures join together? What is the mechanism for forming a chemical bond?

Let us examine the hydrogen molecule-ion. This is formed of two protons (the nuclei) and an electron equally attracted, because of Coulomb forces, by both (Fig. 8).

The electron as a glue

Without the electron the two nuclei, both positively charged, would repel each other, and consequently could not lead to the formation of a bound

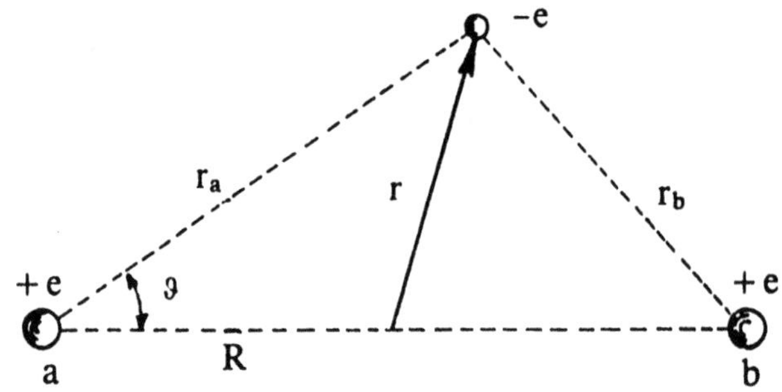

Fig. 8. The hydrogen molecular-ion. The electron, $-e$, is equally attracted, by Coulomb forces, to proton *a* and proton *b*. If the internuclear distance *R* is very large, the electron tends to draw closer to one or the other of the protons, and the system can be described in terms of a hydrogen atom at a very great distance from a bare proton. As *R* decreases, for interprotonic distances comparable to Bohr's diameter of the hydrogen atom, it's as if the electron, unable to decide – so to speak – which of the protons it should approach, opted for both of them and for the central region of the molecule, where, after all, the electrostatic screening action of the repulsive force existing between *a* and *b* is strongest

system. It is the negatively-charged electron therefore that is responsible for the bond between the two nuclei in the molecule. In other words, the electron acts like glue between the two nuclei.

A question arises. Where is the electron?

The electron as a wave

"If you want to know what an electron is doing, forget about it and pretend it's a wave instead. Calculate where the wave is going and that's where you'll find the electron."

This suggestion, formulated by Nevill Mott, is very valuable. The quantum problem of the localization of an electron is traced back to the problem of calculating a wave. Such a calculation may be made using familiar methods from classical mechanics or electromagnetism, and consists in the solution of a wave equation, Schroedinger's equation. The calculation may be avoided however when one is interested in the qualitative aspects.

The electron as an oscillation

As mentioned, it's necessary to substitute the concept of wave, that is of oscillation, for the concept of electron presence. The idea of oscillation evokes the idea of a pendulum. To visualize the oscillation then, let us refer to a simple mechanical model.

In the positions occupied by the two protons (that is, where the electron is most attracted) let us place two pendulums. The electron – whatever its position may be – introduces a coupling between the two protons, reducing the mutual repulsion. To take it into account, let us join the two pendulums to each other by means of a weak spring (Fig. 9).

In order to determine the regions in which the amplitude of oscillation is greater, one must now analyze the movement of the mechanical system thus realized: wherever the oscillation (the wave, that is) is wide, the electron should be found.

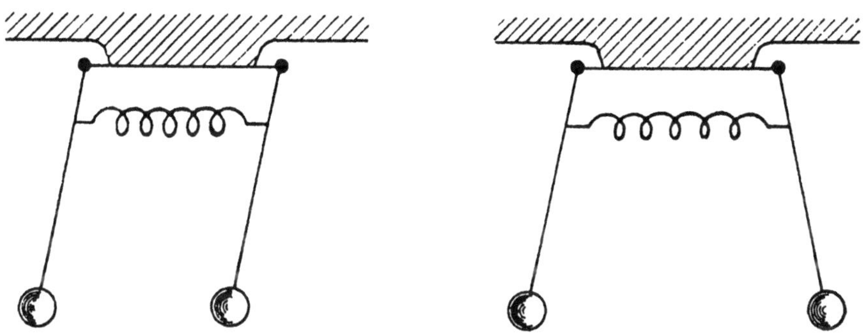

Fig. 9. Characteristic oscillation modes of a system having two pendula coupled by a weak spring. In the *g*erade or symmetric mode, g, the two pendulums oscillate in phase, both with the same frequency, ω_g; in the *u*ngerade or antisymmetric mode, u, the two pendulums both oscillate with the same frequency, ω_u, but out of phase

Vibrational modes of oscillation

This dynamic problem may be solved accurately using the methods of classical mechanics. For this mechanical system there are two typical modes or normal modes of oscillation: the symmetrical mode or *g*erade, ψ_g (Fig. 9g) and the antisymmetrical mode or *u*ngerade, ψ_u (Fig. 9u). Oscillation frequencies ω_g and ω_u correspond, respectively, to these modes.

In the *g*erade mode the two pendulums oscillate in phase, both with frequency ω_g; in the *u*ngerade mode the two pendulums oscillate in phase opposition, both with phase ω_u, slightly higher than ω_g. In the g mode, symmetric, oscillation is on both pendulums, that is, as justice would have it, half on pendulum a to the left and half on pendulum b to the right. The same is true for the u mode, antisymmetric.

Oscillation modes and electron waves

Returning to the hydrogen molecular-ion and to Sir Nevill's prescription for a moment, we must deduce that in state ψ_g, which corresponds to oscillation mode g, the position of the electron is not univocally determined: *the electron is around both protons*, that is, half around a in a state similar to the hydrogenoid state $1s_a$, and half around b. The same is true for oscillation mode u and for the electronic state ψ_u corresponding to it.

A considerable difference exists however between modes g and u. Because of the anti-symmetry of the u mode, interpolating – so to speak – the amplitude of oscillation u, this can be given a null value in the halfway position between the two pendulums: in other words, in state ψ_u, the electron certainly *cannot* be found in the median plane between the two protons. Instead, in state ψ_g, the electron *can* be found in the median plane between the two protons. But it is precisely in the region halfway between the two protons that the screening action of the electron, regarding the electrostatic repulsion between the protons themselves, would be most effective: in state ψ_u, the electron is less bound and less binding than in state ψ_g.

State ψ_g is therefore the ground state of the hydrogen molecular-ion. It is a stationary state, independent of time, in which, as we've said, the electron

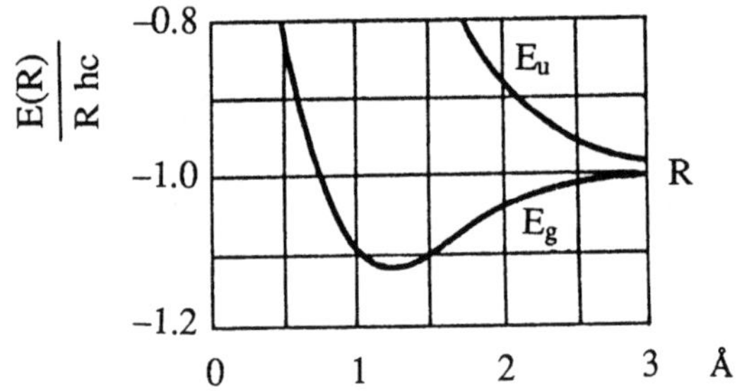

Fig. 10. Dependence of energy levels E_g and E_u of the hydrogen molecular-ion on the internuclear distance R between the two protons. When the distance R is very great, the system's energy corresponds to that of a hydrogen atom in the groundstate energy level $E = -13.6\,\text{eV}$, and is the reference level (this level corresponds to -1.0 Rydberg). With decreasing R, it splits in two. The lower (E_g) refers to the bonding orbital described in first approximation by the wave function ψ_g, and the higher (E_u) refers to the antibonding orbital described by the wave function ψ_u (see Fig. 11)

is *always* half in the neighborhood of a and half in the neighborhood of b, and in particular, it can be found halfway between the two protons. This is a strategic position, in which the electron can enjoy the coulombian attraction of both protons to a greater extent, and thus may exert its bonding function in the molecule more effectively than at any other point.

In forming the chemical bond, the energy aspect – especially the distribution of the energy levels associated to the molecular states – is of obvious importance.

This is the right moment to discuss it.

Electron energy levels and electron states

We should start by saying that in passing from the mechanical model of the double pendulum to the biatomic molecule, the analogue of the oscillation frequency (ω_g or ω_u) is the energy of the levels associated to states ψ_g or ψ_u, that is E_g and E_u respectively (Fig. 10).

If the distance R between the two protons is great, the electron tends to move nearer to one or the other of the protons, in a configuration described in probabilistic terms by the hydrogenoid wave function $1s_a \equiv \psi_b$ or $1s_b \equiv \psi_b$ (see Fig. 6 and 7). In this case, the energy of the entire system is equal to that E_1 of the energy level associated to the ground state of the hydrogen atom: the other proton is at a very great distance, and the perturbation it induces in the hydrogen atom may be ignored. When R decreases however, assuming values comparable to the size of the atomic orbital associated to the ground state energy level (actually a few Angström), the system undergoes radical reorganization: a new possibility opens up for the electron, that of moving nearer to the other proton. The system's original energy level, E_1, splits up into two levels: E_u, above, and E_g, below E_1 (Fig. 10). Correspondingly, the original $1s$ atomic orbitals combine linearly to create two *molecular orbitals*. The first of these, ψ_u, associated to the upper energy level, is called *antibonding* and the second, ψ_g, associated to the lower energy level, is called *bonding* (Fig. 11).

Molecular orbitals

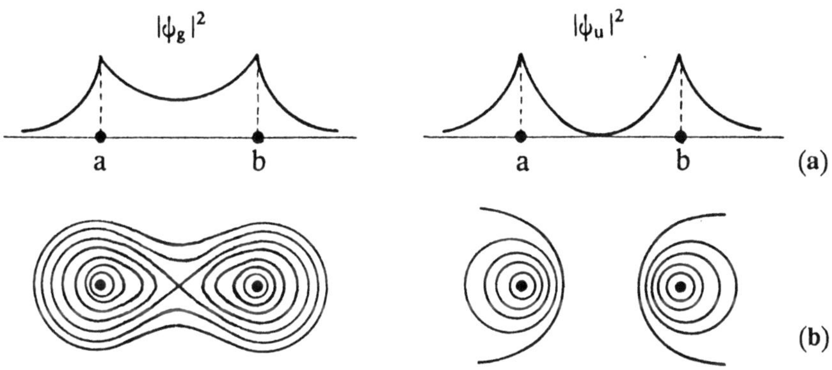

Fig. 11. (**a**) Bonding and antibonding molecular orbitals of the hydrogen molecular-ion. For both figures, the distribution of electron probability in the unit volume within the orbitals, $|\psi_g|^2$ and $|\psi_u|^2$ respectively, is depicted in correspondence to an internuclear separation R = 1.06 A with respect to the ground state of the molecule. Like the contour lines in a military map, the curves (**b**) represent lines of equal electron probability density in a section of the orbital containing the molecular axis. For the bonding orbital, the probability density is relatively high in the molecule's central region. The antibonding orbital instead is virtually cancelled out in the "nodal" plane of symmetry orthogonal to the molecular axis

Probability density

Analogous to the description utilized for the atom, the molecular orbital is a spatiai region bound by a surface on which the state function of the electron responsible for the bond in the molecule is constant; inside the molecular orbital the square modulus of the wave function – and with it the electron probability density – is relatively high; for this reason, the molccular orbital is a space domain inside which one can be reasonably sure of finding the electron. Energy E_g belongs to the bonding orbital; this is inferior both to energy E_u associated to the antibonding orbital and to energy E_1, associated to the non-interacting structural moduli (that is, placed at great distances compared to the "radius" of the atom). Consequently, the bonding orbital turns out to be the one that lends itself best to hosting the electron that is shared by the two protons: in the bonding orbital the electron forms a *covalent* bond between the protons. In the bonding configuration, the electrical repulsion between the two protons of the atomic nuclei, both positively charged and separated, at equilibrium, by a distance of $1.06 \cdot 10^{-8}$ cm, is effectively screened by the electron's negative charge, so that the energy of the system is minimized. In the bonding orbital the electron is in static equilibrium, that is, *always* positioned half on each of the two protons, or on each of the modular elements which form the molecular structure.

Covalent bond

Energy levels of the bonding and antibonding orbitals

As we've said, probability runs high that the electron described by the bonding orbital ψ_g be found in the space region between protons a and b – that is, precisely where the cohesive action resulting from the screening exerted by the electron itself of the electrostatic repulsion between a and b is most effective. On the other hand, for the electron described by the antibonding orbital ψ_u, there is a high probablity that it be found only outside of the same space region: for example, it certainly cannot be present in the nodal area formed by the system's plane of symmetry, where the wave function ψ_u vanishes. This is why the energy level E_g associated to the bonding orbital is lower than energy level E_u, associated to the antibonding orbital.

In the following paragraph we will see how application of an electromagnetic field may perturb the hydrogen molecular-ion, breaking its symmetry and thus dynamically altering its behavior.

3. Binary Structures: Dynamics of Electrons in Molecules

A molecule in vacuum

In the preceding section state functions and energy levels were illustrated for a hydrogen molecular-ion placed in the vacuum, or at any rate in an environment ideally free of radiation having a frequency high enough to connect, spectroscopically, the energy levels associated to the bonding and antibonding orbitals. If we closed the discussion on the covalent bond in molecules at this point, we would give the impression that atomic moduli, combining with each other in the structures, really acquire that configuration of imperturbable stability that calls to mind the soporific " ... and they lived happily ever after" from the fairy tales of by-gone times. Within certain limits, these unique characteristics are actually realized, at least regarding the hydrogen molecular-ion. For the electrons bound covalently in the bonding orbital, the alternative of the antibonding orbital is apparently precluded: in fact, the energy that is on the average associated to a photon of thermal radiation (that is, the product $K_B \cdot T$ of Boltzmann's constant $K_B = 1.38 \cdot 10^{-16}\,\mathrm{erg\,K^{-1}}$ by temperature T expressed in Kelvin degrees) at ordinary temperatures (300 K) is equal to approximately 25 meV; this thermal energy is too low to activate the electronic jump from the lower energy level E_g to the higher energy level E_u: these levels are separated from each other by the energy barrier $E_u - E_g$ of several eV, which seems, like the summit of Mount Everest to a sedentary reader, too high to enable the bonding and antibonding molecular orbitals associated to them to mix (Fig. 10).

A molecule in a resonant electromagnetic field

But what happens in other binary structures, such as the ammonia molecule, in which the difference $E_u - E_g$ is comparable or even inferior to the thermal energy? What would happen to the hydrogen molecular-ion itself if it were placed, instead of in the vacuum, in an electromagnetic field whose frequency corresponds to that of the Bohr transition between E_g and E_u?

On the basis of these queries we have formulated a problem typical of the spectroscopic analysis of molecular structures. We will try to reply by utilizing Mott's prescription once again – that is, falling back on the analogy between electron presence and oscillation of the mechanical system formed of the two coupled pendulums (see previous section).

The dynamics of ambiguity

Let us mention one result immediately (Fig. 12). In the presence of an electromagnetic field of radiation whose frequency corresponds precisely to

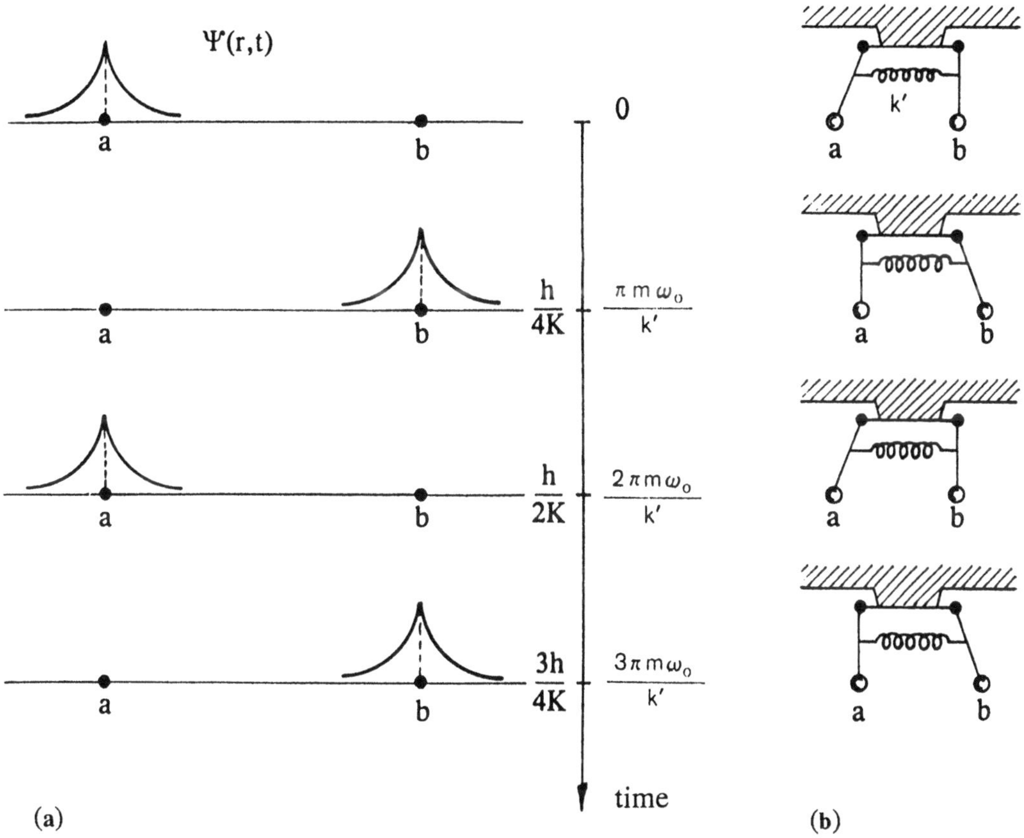

Fig. 12. (**a**) The non-stationary state $\psi(r,t)$ of the electron of the hydrogen molecular-ion. When the proper radiation is present, capable of breaking the symmetry of parity (see Chapter 3) and of connecting bonding ψ_g and antibonding ψ_u stationary states, the new situation may be described by a linear combination, of the type $\psi_g + \psi_u$ or $\psi_g - \psi_u$, of the stationary states themselves. It describes an electron that moves periodically, with time, from hydrogenoid state ψ_a to state ψ_b and vice versa, with a frequency equal, in units of h, to twice the exchange integral, $-2K \equiv E_u - E_g$. (**b**) Oscillation amplitude in a double coupled pendulum. The oscillation shifts periodically, with time, from pendulum a to pendulum b and vice versa, with a frequency determined by the spring that joins the two pendulums. The density of electron *presence*, $|\psi(r,t)|^2$, and the squared modulus of the *oscillation* amplitude of the coupled pendulum may be placed in correspondence (From G. Caglioti, op. cit.)

that of the energy difference $E_u - E_g$, the alternative of gaining entrance to the antisymmetric molecular orbital ψ_u is no longer precluded for an electron initially described by the symmetric orbital ψ_g: more or less what happens when we find ourselves facing the dilemma brought on when a possibility of choice is revealed, and indecision and tensions are created which lead to the dynamics typical of ambiguity.

Stationary states ψ_g and ψ_u, which being characterized by different symmetries could not be mixed in the absence of radiation, are mixed and thus

Combination of stationary states induced by a resonant radiation

define states – $(\psi_g+\psi_u)$ and $(\psi_g-\psi_u)$ – no longer symmetrical and stationary. For an electron described by one of these two states, it's as if it were periodically exchanged between proton a and proton b, with a so-called resonance or exchange frequency equal to that, $(E_g - E_u)/h$, of the resonant radiation. In other words for an electron described by state $(\psi_g + \psi_u)$ for example, it's as if the electron, originally near proton a in hydrogenoid state ψ_a, in time $h/2(E_u - E_g)$ was transferred close to proton b in hydrogenoid state ψ_b, and afterwards returned to state ψ_a again, at time $h/(E_u - E_g)$.

Symmetry breaking

It should be emphasized that this resonant behavior of the electron can be triggered only by something acting externally on the molecule that is capable of breaking the symmetry.

From a stationary to a resonant state

To better understand the mechanisms which, at the moment symmetry is broken, transfer the electron from a stationary, centersymmetric state to a resonant state, let us refer to the double pendulum utilized in the preceding paragraph.

Centersymmetric oscillation modes g and u (Fig. 9), while typical of the dynamics of this mechanical structure, don't exhaust the variety of possible modes of motion. Formally, in fact, each superposition or linear combination of g and u represents a possible movement for the structure, and vice versa. Nevertheless, it makes sense to refer these linear combinations to actual situations of practical interest only when an external action (or, at the atomic level, a radiation) capable of breaking the system's symmetry, is present.

As a matter of fact, only if the symmetry is broken – by triggering at the beginning the oscillation of pendulum a alone, for example – is it possible to couple modes g and u, that is, the centersymmetrical stationary states ψ_g and ψ_u.

The resonant exchange of the electron and the beating in the viola d'amore

Among the possible linear combinations of g and u, the $g + u$ and $g - u$ types are particularly important: these define non-stationary and non-centersymmetric modes, which no longer have a constant frequency. In the state of motion $g + u$ $(g - u)$, oscillation is initially limited to pendulum a (b); it then tends to extend to both pendulums; but after a period of time – the brevity of which increases with the rigidity of the coupling spring – the oscillation passes entirely onto atom b (a), only to return to a (b) again, and so on. This resonant exchange of oscillation between a and b takes the name of beating. It is well-known to those who, while playing the viola d'amore, produce resonance *per consensum* of the sympathetic chords.

As already mentioned, on the atomic scale a spectroscopic line, or rather a resonant electromagnetic radiation, associated to the periodic transfer of the electronic charge from one atom to the other, corresponds to the mechanical beat.

The results we've just established for the dynamics of the electron in the hydrogen molecular-ion should be taken cum grano salis, since this molecule in state ψ_u is unstable and tends to dissociate. Nevertheless, these results can be applied to the hydrogen molecule or extended to other binary structures.

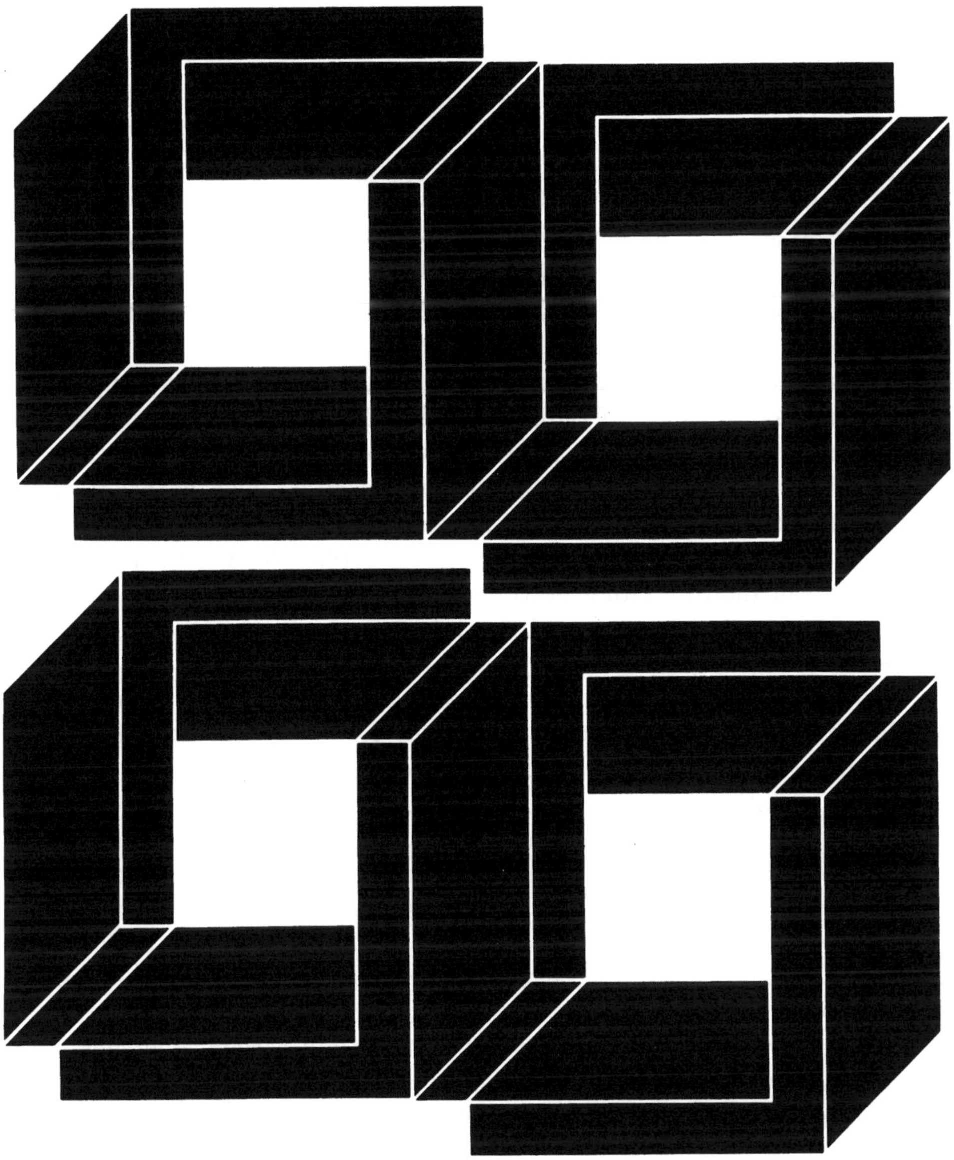

Plate III. F. Grignani, Psychoplastic Structure (acrylic, $70 \times 70\,cm^2$, 1968), details

The ammonia molecule

We're thinking of the ammonia molecule, for example. This molecule has a structure like an umbrella that may be turned upside-down 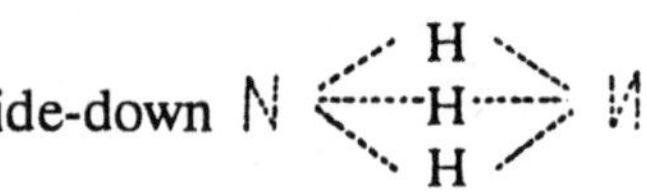 [1].

In this structure, it's as if the nitrogen were undecided between the position to the left or that to the right of the plane of the three hydrogen atoms (Fig. 13): moreover, the difference between the levels $E_u - E_g$ pertaining to the nitrogen's symmetric and antisymmetric states is small (about one ten-thousandth of an electronvolt) with respect to the thermal energy $K_B T$ (about 25 meV).

We could refer as well to certain molecules of coloring agents subjected to an electromagnetic field of radiation having the proper frequency. For these molecules – as will become clearer further on – the energy jump $E_u - E_g$ can be filled in, instead of thermically, only spectroscopically, that is, using a resonant electromagnetic radiation whose frequency ν, multiplied by Planck's constant, is exactly equal to the jump (cfr. the caption for Fig. 6). The resonant radiation can then be either absorbed or emitted by the molecule, and give rise to spectroscopic effects such as those responsible for the stimuli which generate chromatic sensations.

Charge transfer spectrum

For the sake of simplicity, in the following we will refer to the hydrogen molecule. In this molecule the two valence electrons are normally found in the bonding molecular orbital pertaining to the groundstate level, E_g. When resonant radiation is present, one of the two electrons of the molecule may be stimulated to jump to the energy level E_u. At that point it is as if *during* absorption (followed by re-emission) of radiation the wave function of this electron were actually described by a linear combination of stationary states such as $\psi_g + \psi_u$ (or $\psi_g - \psi_u$): the electron, struck by the incident radiation, resonates as previously described between proton a and proton b of the molecule. This constitutes the *charge transfer spectrum* (Herzberg 1950).

Resonance in the chemical bond

From this point on therefore, when discussing certain analogies among resonance in the chemical bond, the dynamics of the perception of ambiguous figures and the acquiring of information in spectroscopic measurements, we

[1] Using dotted lines for bonds or symbols of atomic elements to indicate that they should be considered only half realized, in a structure that is usually depicted like this: 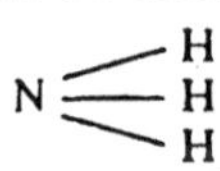, may seem an affectation. In fact, in an ammonia molecule in its stationary ground state, the position of the nitrogen is strictly speaking frozen half to the left and half to the right of the plane defined by the three hydrogen atoms: the ammonia molecule actually has no electric dipole moment, and the value $1.14 \cdot 10^{-18}$ electrostatic units (10^{-18} e.u. = 1 debye = $3.33 \cdot 10^{-30}$ coulomb · meter) that the tables provide "measures the dipole moment associated to the piramidal configuration of the molecule NH_3, which would exist if this configuration were static" (cfr. Pantell and Puthoff, 1969).

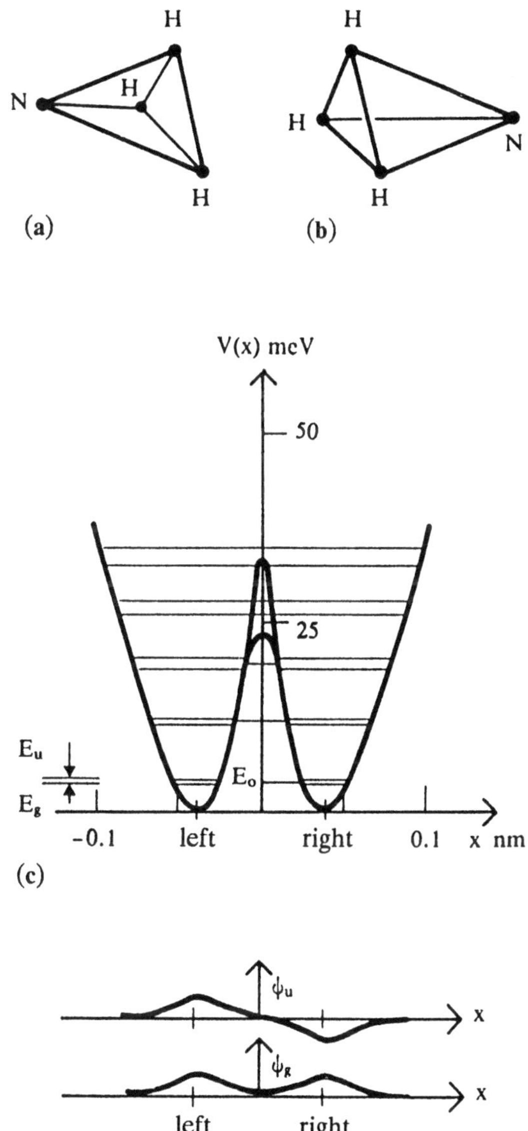

(a) (b) (c) (d)

Fig. 13. The ammonia molecule (**a**) NH_3, or (**b**) H_3N. (**c**) represents the potential energy function for nitrogen along the axis of the molecule, on both sides of the plane defined by the three hydrogen atoms, as well as the energy levels E_g and E_u associated to stationary states ψ_g and ψ_u. (**d**) illustrates the amplitudes of the probability density for the nitrogen atom along the molecular axis in stationary states ψ_g and ψ_u. Along with their square modulus, $|\psi_g|^2$ and $|\psi_u|^2$ provide a measurement of the probability that nitrogen, described by states ψ_g and ψ_u respectively, be found near any given point on the axis of the molecule. State ψ_g, *g*erade, is symmetric, and state ψ_u, *u*ngerade, is antisymmetric. Nevertheless, in both states, the nitrogen presence probability is, always, half to the left and half to the right of the plane defined by the three hydrogen atoms will be referring to one of the two electrons of the hydrogen molecule, instead of the electron of the hydrogen molecular-ion.

The dynamic characteristics of binary structures subjected to a resonant electromagnetic field can be effectively grasped by observing what could be called cis- or trans-perspectives of two identical moduli placed side by side. We are alluding to the dynamic effects which ambiguous figures suggest at the level of perception.

Dynamic perception of ambiguous figures

Just as one can conceive that the generation of the hydrogen molecule initiates with the *atoms* that compose it because of their moving toward each other and the consequent "pooling" of the valence electrons in the covalent bond, so one may obtain an ambiguous figure by bringing the *structural moduli* that compose it into contact, so that they have one part in common. For example, in the graphic representation in Plate IV, the adjacent walls of the two cubic moduli, which then become one, represent the analogue for the electron in the hydrogen molecule. By its construction, the central, inside wall belongs always, stationarily, half to the modulus on the left and half to the one on the right. But, depending on how it is perceived, it seems to belong first to one modulus, then to the other, and then to the first again, *resonating* between the two according to a dynamic alternation of inversions of the perspective: the fascination of ambiguous figures is most likely due to these dynamic effects, a result of the tensions created by the contradictions inherent in their inconcretizable structure.

In bringing this section to a close, we should warn that the analogy, merely touched on here, between the spectroscopic resonance of the bonding electrons in a molecule and the dynamics of the perception of ambiguous figures is apparently pertinent only to the final stage of the process: the behavior of the ambiguous figure visually thought of by the observer is analogous to that of a single molecule subjected to an electromagnetic field, only on the other side of a threshold of the control parameter, beyond which – once a dynamic instability is overcome – the observer-figure organizes itself coherently in a dynamic, ordered structure.

And finally let us add that the resonance frequency with which the shared wall is exchanged between the two cubic modules depends not only on the morphologic interactions among the graphic signs, which the designer imprints in the figure, but on the observer as well: in the last analysis, it is the observer who performs, neurophysiopsychologically, these exchanges.

This discussion will continue in the following section, and will be taken up again in the third and penultimate chapters.

4. Binary Structures and the Bit of Information

Information

This section is dedicated to a preliminary examination of the concept of information. Along with matter and energy, information is one of the fundamental sectors of science and engineering. It assumes an essential role in the evolutive processes of all systems formed of interacting agents, including biological systems and organized societies.

Entropy and information, neg-entropy

The concept of information is delicate. It has stirred controversy even at the level of definition: information as *entropy* (Shannon, von Neumann, 1948); information as the *removal of uncertainty* or the *reduction of entropy*, or *neg-entropy* (Brillouin, 1956); stored information as information understood in Brillouin's interpretation or as redundancy according to Shannon's interpretation, as *capacity to store and transmit significance or knowledge* (Gatlin, 1972); information and entropy as *two completely different concepts linked to each other only by a formal relationship of a mathematical nature* (Fast, 1970); information as a structure's *degree of order* (Monod, 1970). We will return to some of these definitions in the fourth chapter.

Order

Graphic condensation

Let us look again (Plate IV) at the process of "graphic condensation" of the two moduli, separated at the start, which converge in Grignani's binary structure. These moduli are transformed into each other as a result of the inversion center symmetry operation.

Inversion symmetry

The center of inversion symmetry is that point, at the center of each pair of moduli, where a reversal of all the elements of the pair in relation to the point, would be unnoticeable.

Inversion symmetry persists in the system of the two cubic moduli even when they are brought into contact and thus, so to speak, have the possibility of interacting through the common wall.

At this point, as mentioned in the preceding section, one notices however a disconcerting fact: if the two united moduli are observed long enough, all at once a doubt seems to arise: which of the two moduli does the central inside wall belong to? Just as for undecided judges in a dispute between neighbors, the removal of uncertainty consists in affirming: the inner wall belongs to the module on the *left*, or the inner wall belongs to the module on the *right*. The removal of uncertainty coincides with the information associated to the order given the structure. (There is a difference of sign in Brillouin's and Shannon's definitions of information: Shannon's redundancy, or removal or reduction of informational entropy *coincides* with Brillouin's information).

The bit of information

In the binary structure in Plate IV the increase of information corresponding to the decrease of original uncertainty S between the two possibilities, left s and right d, can be considered unity and is equal to a *bit* (*bi*nary dig*it*): that is, a bit is all that's needed to choose: d or s.

Uncertainty as informational entropy

In a more complex system, composed of two binary structures, such as that considered in Plate III, the amount of *uncertainty* (or potential information or informational entropy) incorporated initially in the two structures is double, and the number W of possibilities is double as well. That is, there may be four possibilities or *configurations*: ss (the elements shared between the moduli of the first structure and the second structure both move to the left), sd (element in common between the moduli of the first structure on the left and between the moduli of the second structure on the right), ds and dd. Consequently,

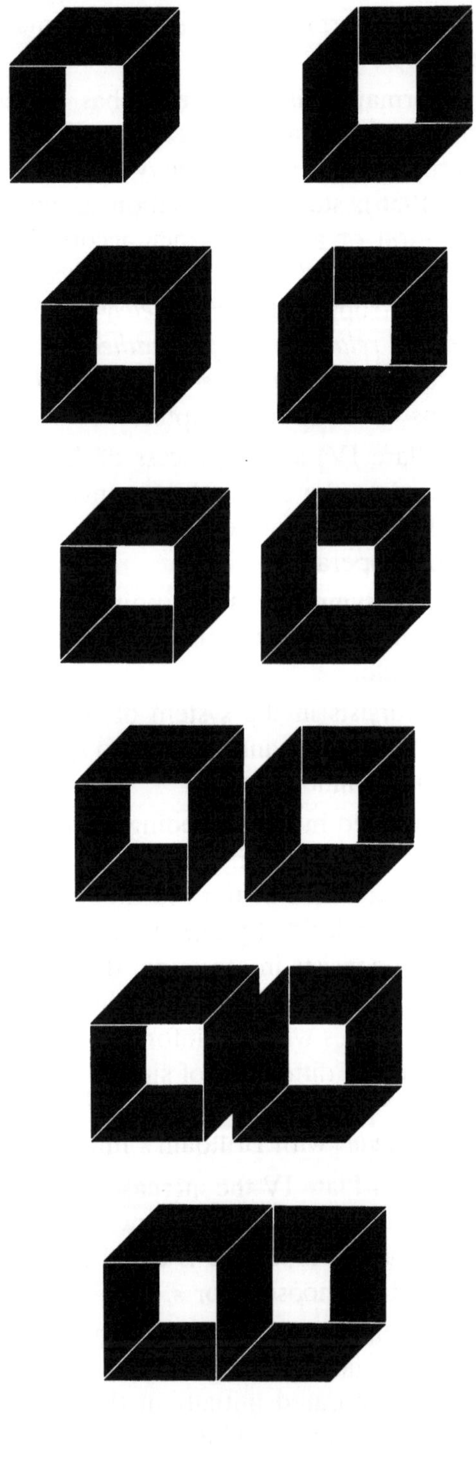

Plate IV. F. Grignani, Graphic Condensation

double the information may be extracted from the system, and two bits are needed to establish its configuration.

It is natural to expect, at this point, that the amount of total uncertainty S incorporated in a system formed of three binary structures be triple that pertaining to a single binary structure. A problem arises however, due to the fact that there are now eight possible configurations:

III bit	↓	II bit	I bit
sss;	*sds;*	*dss, dds;*	*ssd, sdd, dsd, ddd,*

But actually three bits are sufficient for eliminating any uncertainty. For example, if one opts for the second configuration, only one bit is needed for excluding the last four configurations; for excluding the third and the fourth of the four remaining configurations, a second bit is necessary, and a third for disposing of the first of the remaining two.

Anyone who still has doubts should put himself in the shoes of the person who has to organize the eighth, quarter and semi-finals and finals of a tournament (of tennis, for example).

In general, it's expected that

- the amount of uncertainty S incorporated in a structure will increase with the number of configurations;
- for a complex structure, formed by the aggregation of two or more structures, the total uncertainty S be equal to the *sum* of the uncertainties of the individual structures;

Sum of the uncertainties

- the number W of configurations of the complex structure increase with the *product* of the number of configurations of the individual structures that shape it.

Product of the number of configurations

What has been said up to this point regarding the relation between uncertainty S expressed in number of binary choices or bits, and the number of configurations, W, that these choices may be operated on in order to reduce uncertainty to zero, is shown in the following table:

Uncertainty S (bit)	Number W of configurations
1	2
2	4
3	8
4	16
⋮	⋮
n	2^n
⋮	⋮

One deduces that uncertainty increases according to a linear progression as the number of configurations increases according to a geometric progression. In other words, the number of choices needed to remove uncertainty and produce information grows linearly with the exponential growth of the number of configurations on which the choices must be made. The bond between S and W, to be quite clear, is the same type as the one existing between the passing years and the monetary value of goods, progressively swollen by inflation. The behavior is logarithmic, and may be expressed as follows (Fig. 14):

$$S = \log_2 W.$$

For example, in a binary structure where $W = 2$ (right or left d or s), uncertainty or informational entropy S is equal to one ($S = \log_2 2 = 1$). The removal of uncertainty consists in ordering the structure by giving it configuration s (or d). The removal of uncertainty in this elementary example is total. It accompanies a reduction in entropy from its starting value $S_{\text{initial}} = \log_2 2 = 1$ to the final value $S_{\text{final}} = \log_2 1 = 0$. That is, there is an entropic variation ΔS equal to

$$\Delta S = S_{\text{final}} - S_{\text{initial}} = -1.$$

The information, contained in latent or potential form in the initial indifferent configuration, is stored, in actual form, in the final ordered configuration. The negative increase ΔS of entropy at the removal of uncertainty fosters the increase of order in the final configuration:

$$\Delta I = -(S_{\text{final}} - S_{\text{initial}}) = +1.$$

In general

$$\Delta I = -\Delta S :$$

Information as negative informational entropy

information (Brillouin's) corresponds to (Shannon's) *negative informational entropy* or *neg-entropy*: everything that is not known goes in uncertainty or disorder or entropy.

The primordial information that may be extracted from the elementary alternative s (or d) represented by the binary structure considered here is the unit of information. As mentioned, this is called a *bit*.

Let us return to the relationship between Shannon's informational entropy and the number of configurations W. It is formally similar to the following expression:

$$S = K_B \log_e W.$$

This important thermodynamic relationship was suggested by Boltzmann. It defines entropy S, that is, the configurational disorder or uncertainty of a chemical-physical system in equilibrium, in terms of the number of the system's own microscopic configurations W. K_B is the Boltzmann constant, and the base chosen for the logarithmic function is Euler's natural number ($e = 2.71828\ldots$).

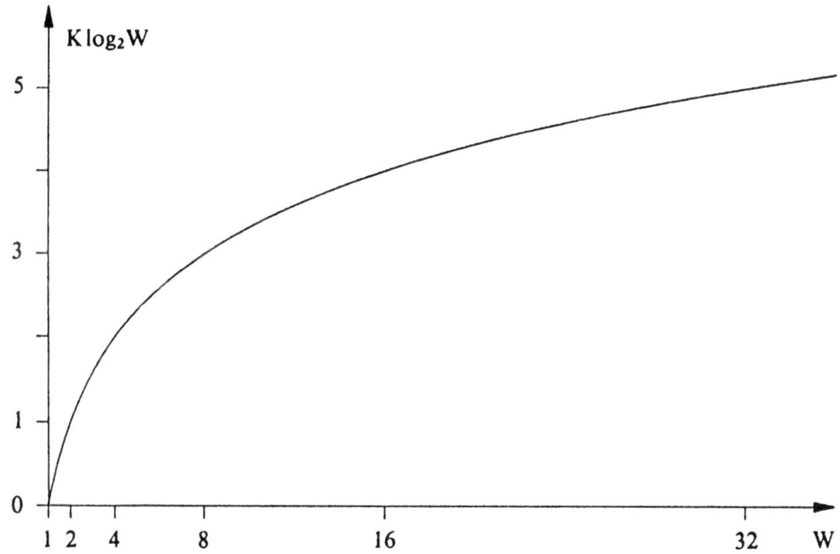

Fig. 14. Behaviour of the total uncertainty S – or entropy – of a system, as a function of the number W of the various microstates or configurations that produce an identical state at the macroscopic level. The inverse dependence of W on S, is similar in nature to that of a sum of money (W) subject to composite interest, which increases exponentially with the passing of the years (S)

Configurational entropy and informational entropy

As we will see in the fourth chapter, Brillouin proposes that the configurational entropy of chemical-physical systems and informational entropy coincide. This identification implies that a quantity equal to (not less than) $K_B \log_e 2 = 0.97 \cdot 10^{-16}\,\mathrm{erg\,K^{-1}}$ corresponds to the bit, that is, to the unit of information or informational entropy.

Energetic cost of one bit

Concretely, according to this point of view, it is necessary to use an energy of at least $K_B T \log_e 2$ to acquire an informational bit, when working at temperature T: the value of this quantity, at room temperature, is decidedly modest: 10^{-27} KWh: possibly, this circumstance was not irrelevant in the development of distributed informatics and, at the same time, of the postindustrial society.

Order and symmetry breaking

We will touch again on the concepts of entropy or disorder, information, order and organization of structures. Nevertheless, with Plate IV still in mind, we should note that, in the very moment in which the structure is *ordered* – inasmuch as the central wall is obliged to opt for the modulus on the left, for example – it appears to undergo a *breaking of the symmetry* of inversion.

Thus it's as if the act of perceiving the structure, or rather of obtaining information about it, provoked a qualitative alteration in the symmetry of the state of the structure itself.

Perception alters the state of an ambiguous figure

From this impossibility of observing an ambiguous structure without disturbing its original state a situation seems to emerge which is similar to that of the spectroscopic observation and measurement of quantum structures. Quantum structures are defined in terms of their stationary states. By definition, the latter are states in which nothing happens, and consequently they are essentially independent of time. But on the other hand they are accessible only by following irreversible procedures which are realized by promoting, over a length of time, spectroscopic transitions which connect the energy levels relating to the stationary states themselves. In other words, spectroscopic observation removes the structure from the stationary state and breaks its symmetry.

Spectroscopic observation alters the state of a quantum structure

Spectroscopic observation, like the perception of an ambiguous structure, is an irreversible process, incompatible with the stationariness of the states in terms of which the structure is described. To grasp, at the intuitive level, the ambiguity inherent in acquiring knowledge of a quantum structure, ordinary language appears inadequate; so-called graphic language however seems to lend itself to this purpose quite well, since it triggers our visual thinking at the instant the ambiguous figure is perceived (see Insert at end of chapter).

Thus there appears to be an analogy between the dynamic behavior of a quantum structure, such as the hydrogen molecule subjected to a resonant electromagnetic radiation capable of bringing on the charge transfer spectrum, and the dynamics of the perception of a dissipative structure, visually thought of as a whole by the observer, beyond a certain threshold of control which the observer himself exercises over the sensorial stimuli.

We will return to this analogy in Chapter Five.

Insert

Symmetry, Information and Ambiguity in Quantum Physics and Design

> There was once a rabbi who was known for his ability to read men's minds and know their thoughts. A boy went to see him and said: "Rabbi, I have a butterfly in my hand. Is it alive or dead?" And to himself he thought 'If he says it's dead, I'll open my hand and let it fly away, on high; if he says it's alive, I'll quickly crush it and show him that it's dead'. Then the boy repeated his question: "Rabbi, I have a butterfly in my hand; is it alive or dead?" The rabbi looked into the boy's eyes and answered calmly: "As you like, my boy, as you like".
>
> Oral Tradition

Description and stationariness, observation, knowledge and irreversibility

The analysis of quantum models of reality, if carried out in ordinary language, the natural means of communication, may lead to overwhelming "logical" contradictions. We are alluding to the problem of spectroscopic observation of quantum structures. And, especially, to the incompatibility between the stationariness of quantum states in whose terms alone these structures may be described, and the irreversibility of the act of observation, needed for acquiring information and knowledge.

Ordinary language and language of art

These contradictions can be avoided however if one agrees to utilize, along with ordinary language, another equally natural means of communication – the language of visual art: a language whose rules – if one may call them rules – are dictated by the necessity of correlating instincts, emotions and illusions; and as such they free us from the rationalistic Aristotelian assumption of the tertium non datur.

Let's take one thing at a time. First of all, let us recall the concepts of state of a quantum system, wave function and quantized or discrete values of the energy associated to possible states. Next we will examine the problem of the "logical" incompatibility between the irreversible process of spectroscopic observation necessary for carrying out the measurement, and the description of a quantum structure by means of its stationary state functions, which depend on time in a way that is in fact physically imperceptible. Lastly, we will suggest an analogy between the process of spectroscopic observation of a quantum structure and the perception of an ambiguous structure.

A wave function is associated to the state of a quantum system – that is to the whole body of experimentally accessible information about the system attributing two "wave functions or stationary states" and two "energy levels" to it analogous to those used in describing the position of the nitrogen in the molecule itself. At first the figure looks like a centersymmetric bi-dimensional design: if we turn each element of the design upside-down with respect to the central point of the figure itself, we're not even aware of having done so. Turning it upside-down transforms the design into itself, that is, it maintains it unaltered, and is therefore an operation of symmetry.

From uncertainty to information via a symmetry breaking

But what happens if the observer controls, beyond a certain threshold, the stimuli that the figure suggests to him? The figure, interiorized in the brain, all at once seems to come alive, to the point of jumping off the page. A kind of letter Z emerges, tilted and askew, with two white bases. A com·mensurate equilibrium among the structural elements of the figure, and with it the original sym·metry, are broken by the observer. At the same time, the indifference and uncertainty inherent in the composition as seen on the plane of the page are removed, and the observer feels inspired to produce information – information that consists, for example, in the choice: the white base on the right (left) is on the near side (on the far side) of the sheet, or in the equivalent choice: the white base on the left (right) is. To whoever asks if, in the exact moment of choice the interiorized structure is still the bidimensional design frozen in a stationary state or if it is already the tridimensional structure of the Z with the white base on the right on the near side of the page, the answer would be *both*: *tertium non datur*?

Perceptive transition of an ambiguous figure

At the precise instant of this perceptive transition the two aspects of that single ambiguous reality that is the interiorized figure exclude each other mutually, while at the same time co-existing in it: the bi-dimensional centersymmetric structure is described by one of the two stationary states but, at the same time, the symmetry of the structure is broken, the two states combine in visual thinking and information is created which leads to choosing, for example, the white base on the right on the near site of the sheet.

Whatever the chance option has been initially – right or left – afterwards the observer can't help perceiving dynamically a rhythmic alternation of perspective inversions right → left → right, etc.

Perception of ambiguous figures

To sum up, the two centersymmetric stationary states are latent in the bidimensional structure and both are needed to describe it. They are orthogonal to each other; that is, they can be combined only if the observer exercises a sufficiently intense control over the stimuli that the interiorized figure calls up in his mind. It is as if such states, once combined at the moment of the initial choice – right or left – induced by the onset of visual thinking, remained coupled to each other when symmetry is broken, so as to suggest to the perception the dynamic alternation of perspective inversions that so closely recalls the resonant absorption or emission of radiation typical of any spectroscopic transition. Si parva licet componere magnis, it would in short seem efficacious to compare spectroscopic observation of quantum structures, which after all forms the experimental base of our present knowledge of the physical world,

Observation of quantum structures

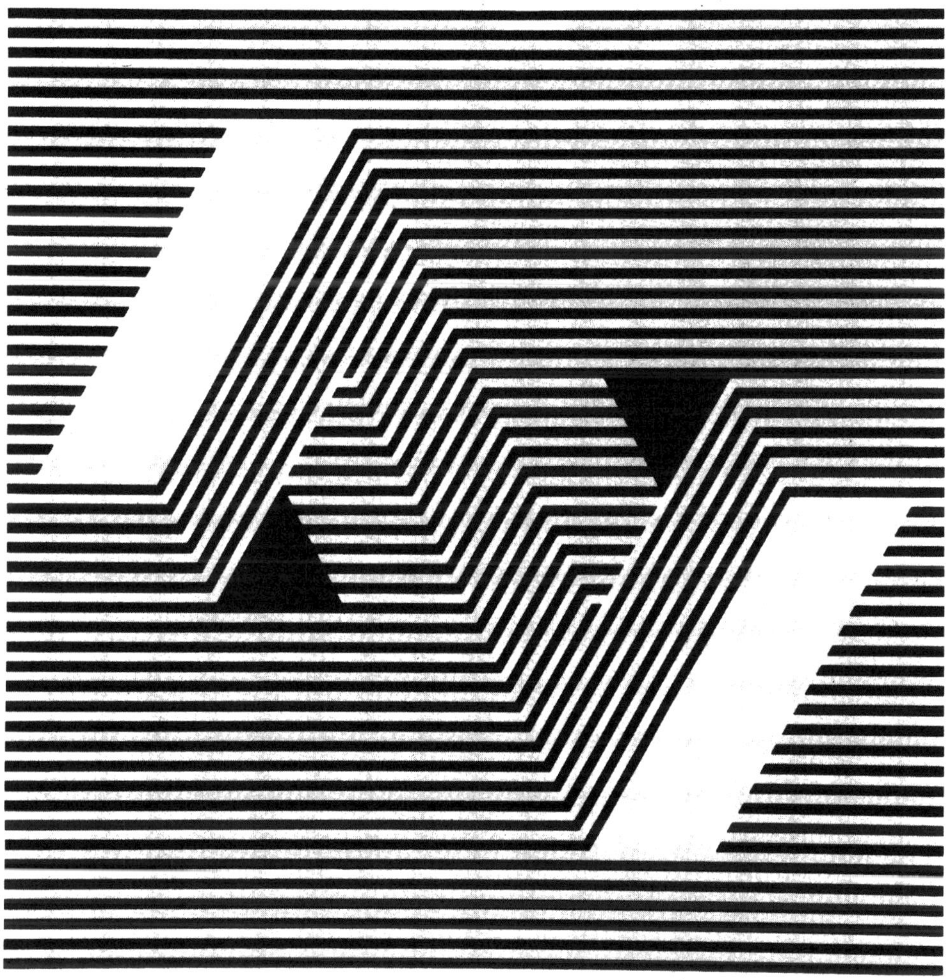

Plate V. F. Grignani, Psychoplastic in Field (406) 70 x 70 cm^2 acrylic, 1971. The perception of ambiguous structures and spectroscopic observation of quantum structures to the perception of ambiguous structures, so disturbing in their undecidability.

This analogy between Grignani's unconcretizable world (Plate VI) and the concrete world that surrounds it, between the emotional fruits of artistic expression and the rational schemes of the scientific method, should of course be taken with a grain of salt. It can nonetheless lead to reflecting on the applicability of a supposition – tertium non datur – that has conditioned western culture profoundly: an assumption whose validity is generally taken for granted in all problems of the so-called exact sciences.

Tertium datur?

Where then is the nitrogen in the ammonia molecule?

Recalling Wittgenstein's warning: "one shouldn't speak of things which can't be spoken about", we could adopt the ambiguous language of silence. But we prefer instead to propose the alternative according to which at times what can't be spoken of can be perceived. In that case we could say, practically at random: the nitrogen is on the right. Or, if you wish, the nitrogen is on the left. Or perhaps, why not, on the right and on the left. The nitrogen, then, can be found where you take it when you look for it.

Thus we've arrived at an elusive conclusion. Or if you prefer, a conclusive elusion. Or perhaps, why not, to both.

Plate VI. F. Grignani, Hyperbolic Tension 1029, 51 × 51 cm^2, Composite Technique

To a superficial observer the work reproduced here could at first seem to be a two-dimensional geometric drawing, composed of many black and white parallelograms, with sharp and angular sides. But after a few seconds these parallelograms reorganize themselves before our eyes: side by side, in pairs, they pile up one on the other in orderly fashion; and like so many symbolic monomers, lined up this way, they appear to depict a system of abstract, polymeric, sindiotactic chains in juxtaposition, which seem to develop along soft and sinuous lines. In exploring the three dimensional space defined by the monomers (similar one to the other) which compose this structure, our glance tends, in other words, to find and subsequently follow preferential itineraries along which the required attention level remains more or less constant. That is, our glance, though alert, tends to stay lazily confined along the polymeric chains among which the attractive morphological interactions which bind the structural modules to each other are most concentrated. After an additional period of observation however, it isn't long before new perspectives explode beyond the surface of this painting. The emotional tension that animates the author, and which the author with a steady hand transfers geometrically into the picture transforms it, as if by magic, into an authentic labyrinth of visual thinking: and it insinuates itself into the observer with the immediacy of direct communication. In allowing his glance to follow a polymeric chain, the observer, at a certain point, distracted by the neighboring chains, relaxes or sharpens his attention to the degree that corresponds to a jump from one polymeric chain to a con-

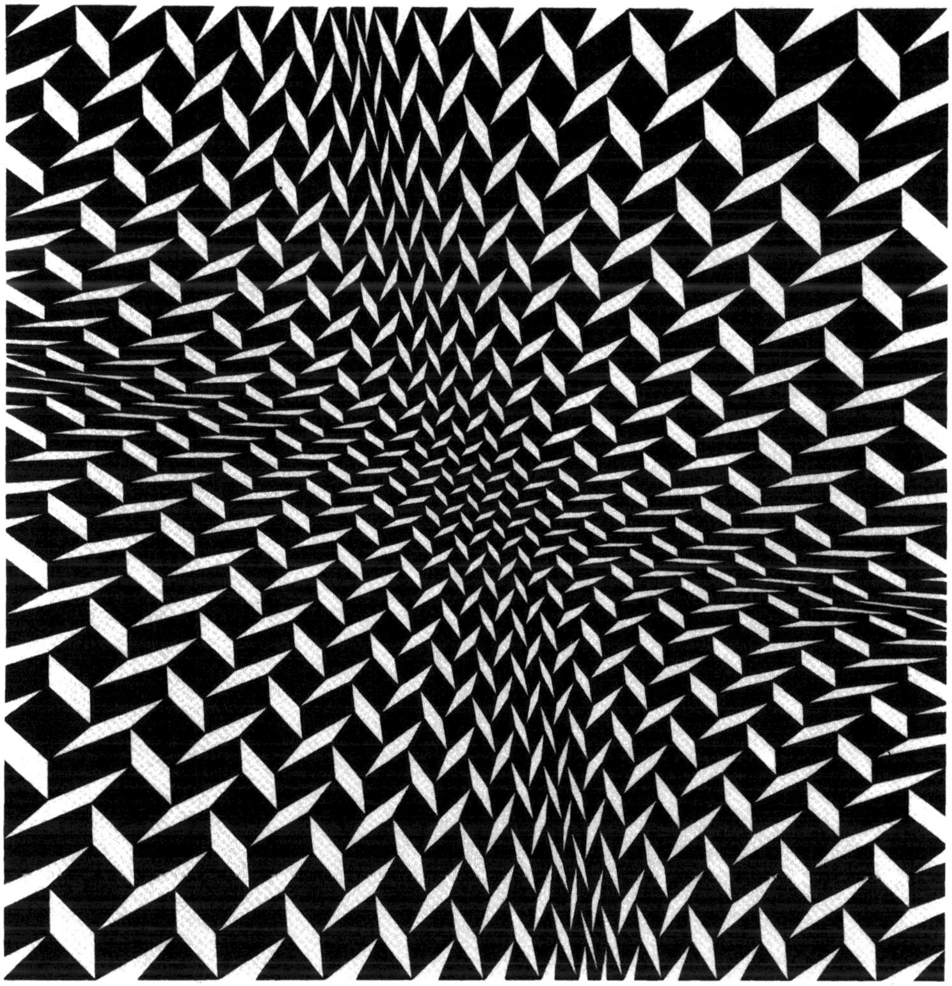

tiguous polymeric chain. Then, all at once, the chains separate from each other and reappear out of phase, only to be substituted by two twisted flights of stairs, which then overturn, interfering with each other. This produces a succession of varied coherent schemes for possible organization of the space, co-existing yet mutually incompatible, unrealizable as an ensemble. Our visual thinking, fascinated, agitated and entrapped among these schemes, flounders oscillating giddily, then it dissolves, as if dazzled, only to return, curious, and recompose itself. Similar to certain pure baroque music, and unlike informational works such as those of M. C. Escher, the *form*, that is the geometric language through which Grignani expresses his tensions rationally, seems to organize itself in signs full of carefully distributed dynamic components, and thus eventually ends up coinciding with the tensions themselves, and therefore with the emotional *content* that has released these tensions. The text suggests connections that might be established between the perceptive process relating to Grignani's gestaltic structures, incredibly laden with unresolvable ambiguity, and fundamental problems of physics. We are only trying to offer here an analytical interpretation of a system of signs which, although rationally organized in a way that appears to conform to the physical-mathematical bases of the logic – still obscure – of graphic language, demand instead to be interpreted coherently, understood synthetically and utilized, above all, emotionally, in their inexpressible and unrealizable whole

Chapter Three

Symmetry and Broken Symmetry in Science, in Perception and in Art

美とはあえて均衡を破ることである。

禅

True beauty is a deliberate, partial breaking of symmetry.

Zen Proverb

Symmetry elements

Symmetry operations

Many structures possess elements of symmetry (mirror planes, rotation axes, inversion center, etc.). They are transformed into themselves as a result of symmetry operations such as a reflection on a plane, a rotation around an axis, an inversion with respect to a point. By definition, such symmetry operations, once made, preserve the original structure. Thus, if we set a ball revolving around its center or one of its diameters, or a glass around its axis, we have no way of knowing that the rotation has been made. That is, it is impossible to measure an absolute, angular coordinate: measuring in fact implies a reference for the angles such as a notch, for example, which cannot be marked on the ball's surface or cut into the rim of the glass without at the same time breaking the rotational symmetry.

Symmetry as indiscernibility

The indiscernibility of a rotation made around an axis or a center of symmetry should be of no importance to us. Indeed, the result of any symmetry operation may seem banal: but often the inconclusiveness which by definition follows on a dynamic action provokes interesting stimuli. We perceive this when our gaze restlessly examines the rose-window of a Romanic church or the facade of a Renaissance building, searching for a reference point, or a confirmation that the architectural work really has no defects and is ideally symmetrical. And it feels rewarded when it finally discerns a structural modulus, an ubi consistam, which unpredictably breaks the architectural symmetry.

An ancient Zen proverb says: "true beauty is a partial, deliberate breaking of symmetry". Pierre Curie suggests "it is dissymmetry which creates the phenomenon".

Actually, the human soul appears to be constantly trying to find an equilibrium between two contrasting and contradictory attitudes. On one hand, the

Repetitiveness

ancestral longing to abandon oneself to a reassuring repetitiveness, like the one that cradles us beating time with repetitive symmetry and always in the same way, whether it be the meter in poetry, the beat of the dance, the litany of prayer or the slogans in the liturgy of political manifestations. On the other hand, as if we were awaiting Godot, the yearning to be involved in exceptional

Dynamics

phenomena or take part in extraordinary events, originating in breakings of symmetry which trigger dynamics that offer us an escape from exasperating routine. This equilibrium can apparently be achieved only rarely, for example in re-living – in the presence of a work of art – the moment of artistic creation. Or listening to Bach: a "great" who, though probably unaware of the Zen definition of beauty, was able in the code of his work to reach an equilibrium, which we would qualify as genetic, between the structural symmetry of musical architecture and order in the choice of the notes of the melodic voices which, horizontally correlated in the counterpoint of the score, become harmony when vertically perceived.

Symmetry in arts and sciences

In the arts, inserting elements of symmetry in a structure is a primordial need of aesthetic character. At the same time, in the field of science, searching for symmetry operations is a required step in the methodology of structural analysis.

The artist who creates a work of art, more or less intentionally tries to bring out its symmetry. Similarly, the scientist who analyzes a structure asks himself, first of all, which operations can be carried out on the structure without causing it to undergo alterations.

The ancient Palazzo del Podestà at Fabriano is centered on an elegant, simple, symmetric main door. On the left however an oblique wall runs toward and intercepts a two-light window, tending to hide it from sight. Vice versa, on the right, the two-light window placed symmetrically with respect to the door's axis is not undermined by any architectural element. So as not to break the symmetry completely, a niche has been carved in the wall at the same distance as the intersection with the two-light window on the left, so that this window and the one symmetrical to it are more or less equally visible.

The need for symmetry

The need for symmetry, so alive in the medioeval architecture of Fabriano, is the same that animates, with a dedication that knows no boundaries nor budget limitations, today's high-energy physicists. They have classified the many so-called elementary particles in a sort of Mendeléef's table, composed – like that of the great Russian chemist – on the basis of symmetry considerations. The particles that haven't yet answered the roll-call however can be created only at the price of a vast amount of energy, which must be made available in the laboratory system in each elementary process of creation. Today the need for symmetry leads to the construction of gigantic accelerators for particle and anti-particle beams, which run in cross directions along a ring eight kilometers in diameter toward a frontal collision. To run this accelerator, 200 MW of power are needed, more or less equivalent to the electrical power supplied for the population of Pavia.

In bringing the chapter on symmetry to a close in one of his treatises, Feynman, after pointing out that an almost total symmetry characterizes not only the structures but even the basic laws themselves that govern behavior of the physical world, asks himself the reason for certain apparent violations of symmetry (among which is the important one leading to the non-conservation of parity, or the non-indifference between right and left in weak interactions, demonstrated experimentally by C. S. Wu and collaborators in 1957).

The need for symmetry breaking

"No one has any idea why. The only thing we might suggest is something like this: there's a gate in Japan, a gate in Nikko, which is sometimes called by the Japanese the most beautiful gate in all Japan; it was built in a time when there was great influence from Chinese art. This gate is very elaborate, with lots of gables and beautiful carving and lots of columns and dragon heads and princes carved into the pillars, and so on. But when one looks closely, he sees that in the elaborate and complex design along one of the pillars, one of the small design elements is carved upside down; otherwise the thing is completely symmetrical. If one asks why this is, the story is that it was carved upside down so that the gods will not be jealous of the perfection of man. So they purposely put an error in there, so that the gods would not be jealous and get angry with human beings".

Reversing the argument, Feynman suggests another reason for breaking symmetry: "God dictated laws of symmetry only in part, so that we wouldn't be jealous of His perfection". But there could also be another explanation for our attitude regarding symmetry: God wants man to be dynamic. He therefore left him free to create order and He ordered him to aspire to freedom: breaking symmetry, following symmetry ideals. As he pleases.

The symmetry and order produced by a breaking of symmetry are two values that belong to science, to perception and to art.

Symmetry returns to the impossibility of being aware of variations caused by transformations: symmetry is what remains unchanged, what is preserved in evolution (entropic). It is possible to cause breakings of symmetry by means of the (neg-entropic) action of external fields, the effect of which is usually that of introducing correlations – or establishing order – among the structural moduli. At the instant symmetry is broken, the structure undergoes conclusively an important dynamic alteration which may be measured spectroscopically or perceived dynamically. In certain situations it is the act of measuring itself – or (as for ambiguous structures) of perceiving – which provokes the breaking of symmetry.

The act of measuring breaks the symmetry

In this chapter the rapport among symmetry, conservation and constants of motion is examined first. Then we deal with the problem of symmetry-breaking brought on by external fields, with particular reference to:

The Stark effect

- the Stark effect caused by an electric field, on the hydrogen atom, for example;
- the hybridization of the atomic orbitals which precedes the formation of molecular structures;
- the perception of ambiguous structures.

Aesthetics and natural laws

The understanding of these delicate matters, aided by examples taken from art through the ages, is facilitated by intuition and even retraced to the emotion stimulated by aesthetic sensations: in this way one may confirm the possibility of grasping, synthetically and at the intuitive level, the essence of the fundamental laws of physics which are usually deduced, starting from experience, using a logical and analytical procedure.

1. Symmetry, Conservation and Constants of Motion: The Privileged Role of the Energy Operator [1])

"The root of all principles of symmetry in physics lies in the assumption that it is impossible to measure certain basic quantities" (Lee, 1968). In a gas, for example, the value of the potential energy is not changed by a rigid translation of the gas in relation to the observer or – the same thing – of the observer in relation to the gas. The potential energy depends only on the relative distance between the atoms of the gas. It follows that a rigid displacement of these atoms does not create any force. The impossibility of perceiving such a displacement or – the same thing – the *impossibility of making a measurement of absolute position* or of determining a privileged origin of the reference system gives the gas *translational symmetry*. By the second principle of dynamics, the cancelling out of the force arising from a rigid displacement of gas leads to the conservation of the total momentum of the gas itself. Translational *symmetry* thus involves the *conservation* of the momentum (Plate VII), that is of the product of the total mass multiplied by the system's ensemble velocity.

Symmetry as conservation

The constants of motion

The momentum generates the translation operator. If the gas has translational symmetry, its momentum is a *constant of motion*: that is, the measurement of the gas' total momentum at a particular instant gives a result which coincides with that obtainable by means of an additional measurement in each successive instant.

---

[1] Prof. Kun Huang of the Academia Sinica, with whom I spoke some years ago, did not hide his skepticism on the possibility of classifying as "popular" a book in which the concept of hamiltonian operator is used. For this reason I preferred using the expression "energy" instead of the expression "hamiltonian" – more familiar, in this context, to those working in the field. On the other hand, as Feynman rightly holds (1965), "Hamilton, who was working in the 1830s, managed (a century later, Editor's note) to associate his name to a quantum mechanics operator, thanks to a quirk of history. It would have been much better to call it energy operator..."

We should add here that a differential operator is formally associated to each of the dynamic observables introduced in this section (energy, momentum, angular momentum, etc.). The differential operator is defined by the fact that, applied to a function, it transforms it into another function. If however the function to which the operator is applied coincides with an eigenfunction of the operator itself, the effect of the operator on that function is simply that of multiplying it by a numerical constant.

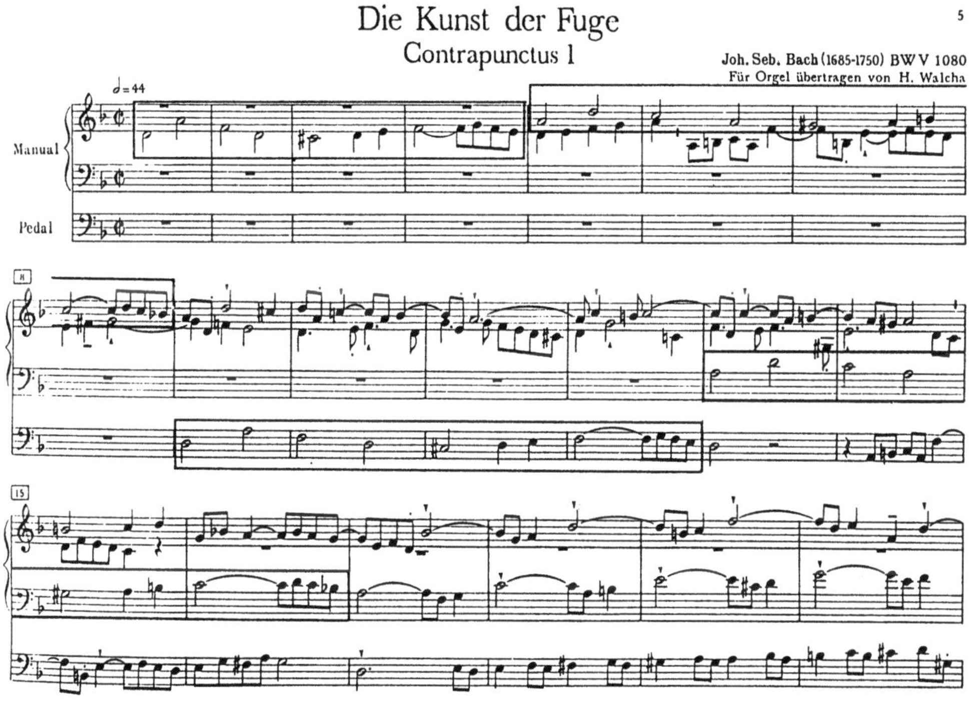

Plate VII. Translational symmetry and the constant momentum (fugue) in the first beats of "The Art of the Fugue", J. S. Bach, BWV 1080

The momentum operator generates the translations. The theme is repeated with translational symmetry every four beats and, running after itself, gives the idea of fugue, that is, of a constant momentum

Homogeneity and conservation of momentum

In short, the homogeneity of space, or the impossibility of realizing that an arbitrary displacement of space itself has been made, leads to the conservation of momentum.

Isotropy and conservation of angular momentum

Similarly, the isotropy of space, or the impossibility of perceiving that an arbitrary rotation of the space itself has occurred, leads to the conservation of the angular momentum. For example, in the Keplerian movement of a planet around the Sun, the conservation of the angular momentum is transferred into the well-known law which states that the area of the sector swept in the unit of time by the planet in the plane of its eliptical orbit – areolar velocity – is a constant of motion.

Inversion symmetry and conservation of parity

Moreover, the impossibility of realizing that a reversal has occurred between left and right, between over here and over there, between above and below (Fig. 15), leads to the conservation of parity: in centersymmetric structures – for which parity is, precisely, a symmetry operation – the structure's state function remains unaltered (*g*erade) or changes sign (*u*ngerade) following an inversion of the system's coordinates with respect to the origin ($\mathbf{r} \rightarrow -\mathbf{r}$): and constants of motion $+1$ or -1, respectively, are associated to the parity operator that has produced this inversion.

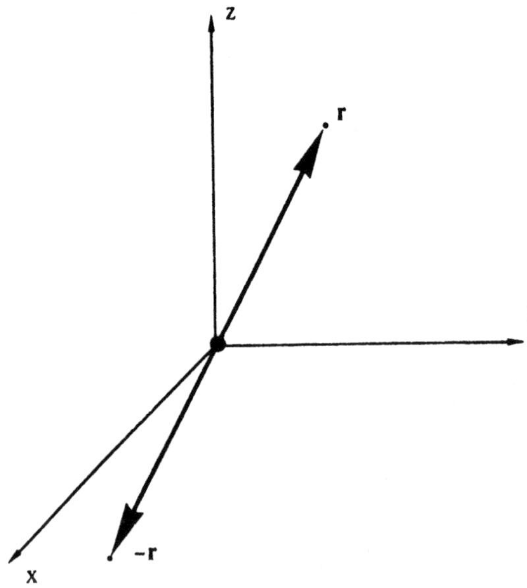

Fig. 15. The parity operator P changes point $+\boldsymbol{r}$ (of coordinates (x, y, z)) into point $-\boldsymbol{r}$ (of coordinates $(-x, -y, -z)$). P is a symmetry operator if it is not possible to perceive that the transformation $\boldsymbol{r} \to -\boldsymbol{r}$ has been made

Homogeneity of time and conservation of energy

Finally, the homogeneity of time, or the impossibility of perceiving that one has made an arbitrary translation of time itself, leads to the conservation of energy. As R. P. Feynman (1965) would say, the "delightful operation which consists in waiting a few moments" is generated by the energy operator. This operator includes "the physics of the situation" and therefore, in particular, it describes any possible alteration that the passing of time brings about in the system.

An isolated quantum structure is characterized by a stationary state in which, by definition, it is impossible to be aware of perceptible changes due to the passing of time. Energy H generates the operator of symmetry translation in time, and in the isolated structure retains its initial value: energy E is a constant of motion. The ground state $1s$ of an isolated hydrogen atom for example is a stationary state in relation to which the electron's way of being around the proton remains the same with the passing of time (Fig. 7) and energy E remains constant – constant of motion – and equal (Fig. 6) to $-Rhc$ (eigenvalue E_1).

In a macroscopic system the energy depends on the coordinates and the momentums of all the material points that make up the system itself; these coordinates and momentums usually vary in a way that is so complex as to be practically indescribable: but the energy, which is their function, remains constant.

Because it remains permanently unaltered, energy is thus, in any case, a necessary reference for understanding the significance of a system. Actually one would be tempted to attribute to the energy of a structure the significance of the structure itself.

Energy and significance

Let us return to the isolated quantum structures.

The role of energy in these structures turns out to be privileged from the formal point of view as well: for two reasons.

The energy operator: its eigenfunctions and eigenvalues

In the first place, the operator H associated to the energy enables us to find the complete system of the structure's state functions – *eigenfunctions of energy* – and the energy levels – *eigenvalues of energy* – associated to them. In principle, once these eigenvalues as a whole are known it's possible, using the methods of statistical mechanics, to find all the measurable thermodynamic quantities which, from the macroscopic point of view, characterize the state and the irreversible evolution not too far from equilibrium of a structure when under the influence of external fields. In other words, once the eigenfunctions and eigenvalues of the energy are known, all the information that it is physically possible to obtain on a quantum structure, becomes accessible. Consequently, it is natural to take energy H as a reference for any observable Ω of the system in which one is interested.

Energy as a reference

We are referring to observables such as:

Angular momentum, parity, momentum

- *the angular momentum*, which generates the rotation (symmetry) transformation;
- *parity*, P, or the right – left, this side – that side, above – below inversion;
- the *momentum*, which generates the translation in space (of symmetry) transformation.

As we'll see, the comparison between the Ω observables listed and H can simplify determination of the eigenfunctions and eigenvalues of H considerably. And this is the second reason for which energy H assumes a privileged role.

But what do we mean by comparison between the observables? What sense does it have to compare non-homogeneous observables? And especially, what does it mean to compare energy with observables such as momentum, parity, the moment of the momentum and so on?

A planet around the Sun

An example is badly needed. As usual, we'll call on the hydrogen atom. But before going any further, the reader should take a look at Fig. 16. A bit earlier we recalled that in the Keplerian movement of a planet around the Sun, both energy H and the areolar velocity or, in the last analysis, the angular momentum, remain constant. The Keplerian motion of the planet is the classical analogue of an electron's movement around the proton: it's natural to ask then whether there's any relation between energy and the angular momentum for the hydrogen atom as well.

An electron around a proton

In the spherically symmetrical Coulomb field generated by the positive charge of the proton, the electron's potential energy actually depends only on the electron-proton distance. This distance remains the same if the electron position is confined to a spheric surface centered on the proton: as the elec-

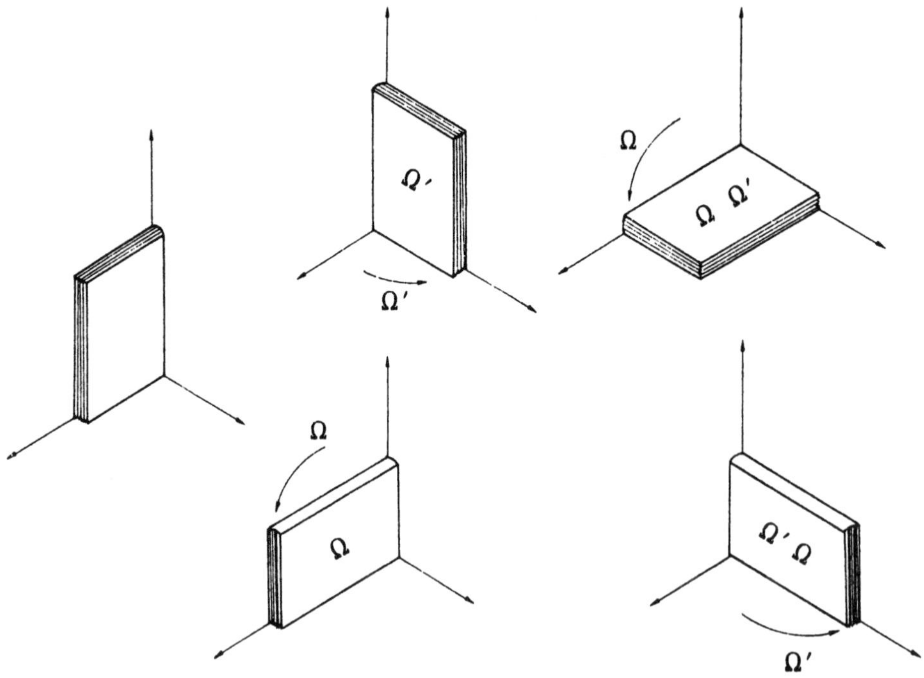

Fig. 16. The comparison between the two observables Ω and Ω' (or between the two operators or the two transformations associated to Ω and Ω') is established through the commutator of Ω and Ω'. The comparison is unproductive if they commute. Two Ω and Ω' transformations do not always commute: applying first operation Ω' and then operation Ω (thus producing operation $\Omega\Omega'$) is not always equivalent to applying first operation Ω and then operation Ω' (producing operation $\Omega'\Omega$): stirring the sugar before pouring coffee in the cup doesn't give the same result as stirring it after. Take a book and stand it vertically on a table. Then turn it counterclockwise 90 degrees *first* around a vertical axis (operation Ω') and *next* around a horizontal axis (operation Ω): as a result of operation $\Omega\Omega'$, the book will appear to be lying flat on the table. Put it back into its original upright position. Starting from this position, apply *first* operation Ω and *next* operation Ω': as a consequence of operation $\Omega'\Omega$, the book will seem to stand up on the vertical plane, resting on its spine. Thus operations Ω and Ω' do not commute, in the sense that operation $\Omega\Omega'$ has a different effect on the book than operation $\Omega'\Omega$ does: $\Omega\Omega' \neq \Omega'\Omega$. If vice versa Ω and Ω' consisted in a rotation around the same axis and in the same direction as two angles equal to 30° and 60° respectively, for example, operations $\Omega\Omega'$ and $\Omega'\Omega$ would both identify with the 90° rotation. In that case, $\Omega\Omega' = \Omega'\Omega$ and operations Ω and Ω' commute

tron rotates on the sphere, its potential energy therefore remains unchanged. In other words, the transformation induced by the electron's rotating does not alter its energy value, and the comparison between rotation and energy is unproductive: rotation and energy commute, so that rotation becomes a symmetry operation. But the fact that, because of the rotation on the sphere, the potential energy does not change, implies that the momentum of the forces acting on the electron is cancelled out. This in its turn implies that the value of the angular momentum of the electron around the proton remains equal to a constant. But every observable which, like energy, remains unaltered with the passing of time, is associated to a symmetry operator. Along with the rotation then, the angular momentum – which generates the rotation – is also associated to a symmetry operator. The values resulting from measurements of the angular momentum operator remain constant in time and belong to the so-called constants of motion.

In general therefore, the comparison between H and each of the Ω transformations which one cannot exclude a priori as a symmetry transformation, may facilitate the search for the eigenfunctions and eigenvalues E of energy H: paradoxically, it is exactly when this comparison is unproductive, that is when Ω commutes with H, that it is effective. In this case in fact the Ω transformations and their generators turn out to be symmetry operations and, as mentioned, define the structure's constants of motion; the eigenfunctions of operator H then are also eigenfunctions of the operators pertaining to them.

Identifying the symmetry operations thus simplifies considerably the search for the set of the structure's state functions.

From the fact that P is a symmetry transformation, for example, it follows that the eigenfunctions pertaining to possible energy states of the hydrogen atom are also eigenfunctions of parity: that is, they may be only the *g*erade type, or symmetric because of an overturning of the electron's coordinates with respect to the proton, or they may be *u*ngerade, that is antisymmetric.

Moreover, the eigenvalues associated to symmetry observables do not change with the passing of time. As we said, these are the constants of motion, and they coincide with the possible results of measurements of the symmetry observables themselves.

Analysis of quantum structures, fruition of art works

The standard procedure for analyzing quantum structures is substantially no different than the process of observing a painting or an architectural work. More or less consciously, in rational analysis just as in artistic fruition, priority is given to searching for the operations or elements of symmetry in the structures (Plate VIII): nor could it be otherwise: symmetry operations, linked as they are to what remains constant in the structure – and in first place to the energy – epitomize the structure's significance and form its essence.

2. Perturbation-Induced Breakings of Symmetry and Restructuring of the Dynamic Architecture of Atoms and Molecules

In the foregoing analysis of the characteristics of symmetry in quantum structures we used the comparison between operators associated to certain transformations (rotations, translations, inversion, etc.) and the energy operator. We have seen that such transformations are symmetry transformations for the structure being studied if comparing them with the energy is unproductive. If symmetry operations are performed on the structure, it remains the same as it was. We have also seen than the so-called constants of motion are associated to symmetry operations and to the dynamic quantities which generate such operations (for example, translational symmetry and constant momentum, respectively).

Energy perturbation and symmetry breaking

What happens to symmetry transformations, to their dynamic generators and to the constants of motion – in other words, what happens to the dynamic architecture of the eigenfunctions and the eigenvalues of a system – when an external energy perturbation, in changing the energy operator, breaks the symmetry of the structure?

One usually has recourse to the theory of groups (Heine, 1960) and the perturbation theory for the reply to such questions. But if we can be satisfied with qualitative answers, it's possible to do without these formal tools (Plate IX).

Whether it involves a quantum structure or a dissipative structure, the breaking of symmetry implies radical restructurations. Breaking symmetry involves the perception of a change in a system originally in a stationary state. For quantum structures, the breaking of symmetry occurs at the moment of observation or measurement. For dissipative structures – we are thinking of the ambiguous structures interiorized during the perceptive process, for example – the breaking of symmetry takes place at the moment of instability, or perception. Thus analogies appear between the processes which cause a breaking of symmetry in quantum structures and in the dynamics of visual thinking.

Symmetry breaking in measurement and perception

In the following, modalities and effects of breakings of symmetry will be considered, both for quantum structures – in the hydrogen atom, in primis – and for interiorized ambiguous figures.

Let us take things one at a time. Applying an external field to a structure modifies its energy operator. The latter in fact must combine the energy of interaction between field and structure. Consequently some dynamic symmetry transformations, for which comparison with the original energy was unproductive may, in a manner of speaking, turn out to be active when compared with the perturbed-energy operator, which includes the energy of interaction. Such transformations then cease to be symmetry operations, and the observables that produce them are removed from their role as constants of motion. The architecture of the perturbed system and its eigenfunctions and their relative eigenvalues as a whole therefore undergo profound changes.

The Stark effect in hydrogen

At the microscopic level, the Stark effect is an example of symmetry breaking in a hydrogen atom brought on by an electric field. In spite of its being a more or less irrelevant spectroscopic effect, we mention it because it enables us to see important breakings of symmetry produced by local interatomic fields in perspective: we are referring to the hybridization phenomena of atomic orbitals which precede the forming of molecular orbitals and, in general, the processes establishing correlations or order among atoms when clusters are formed.

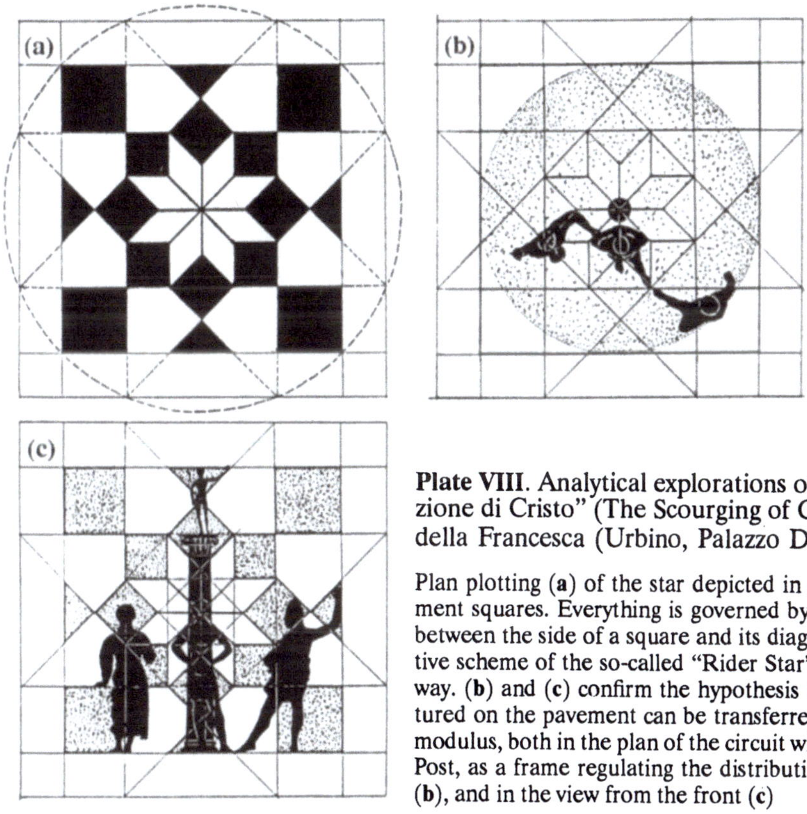

Plate VIII. Analytical explorations of the "Flagellazione di Cristo" (The Scourging of Christ) by Piero della Francesca (Urbino, Palazzo Ducale)

Plan plotting (**a**) of the star depicted in one of the pavement squares. Everything is governed by the relationship between the side of a square and its diagonal. The figurative scheme of the so-called "Rider Star" emerges in this way. (**b**) and (**c**) confirm the hypothesis that the star pictured on the pavement can be transferred, as a figurative modulus, both in the plan of the circuit with the Whipping Post, as a frame regulating the distribution of the figures (**b**), and in the view from the front (**c**)

The graphic reconstruction of the figurative modulus in the form of a rider star – which with its metric scans not only the squares in the pavement, but the whole architectural space in which the characters are distributed in orderly fashion – confronts us with an elegant structure, laden with symmetry. The visual reduction of the model from which the artist drew his inspiration was obtained by Carlo L. Ragghianti and Teresa Zanobini Leoni, with the help of a computer (from Sound Sonda, n. 1, 1 March 1978, Rivista di Studi Elettronici, Università Internazionale dell'Arte, Florence)

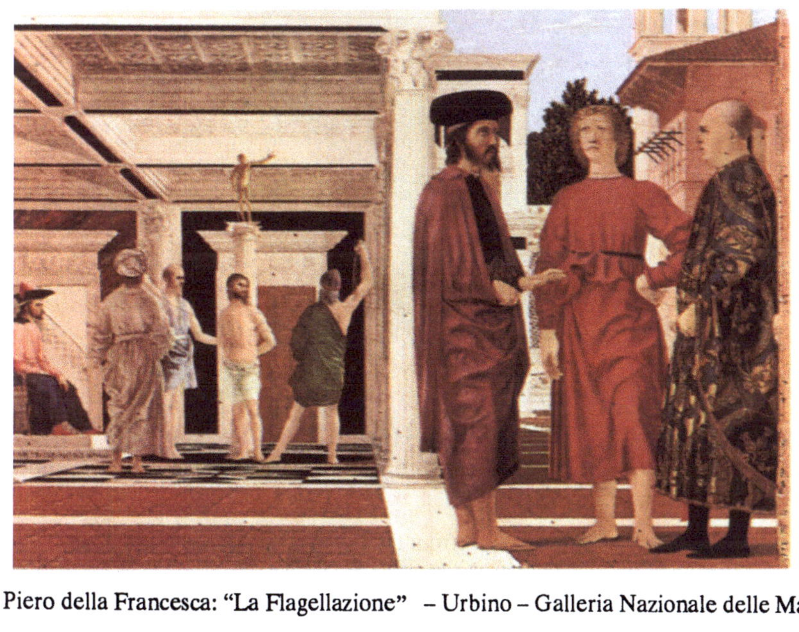

Piero della Francesca: "La Flagellazione" – Urbino – Galleria Nazionale delle Marche

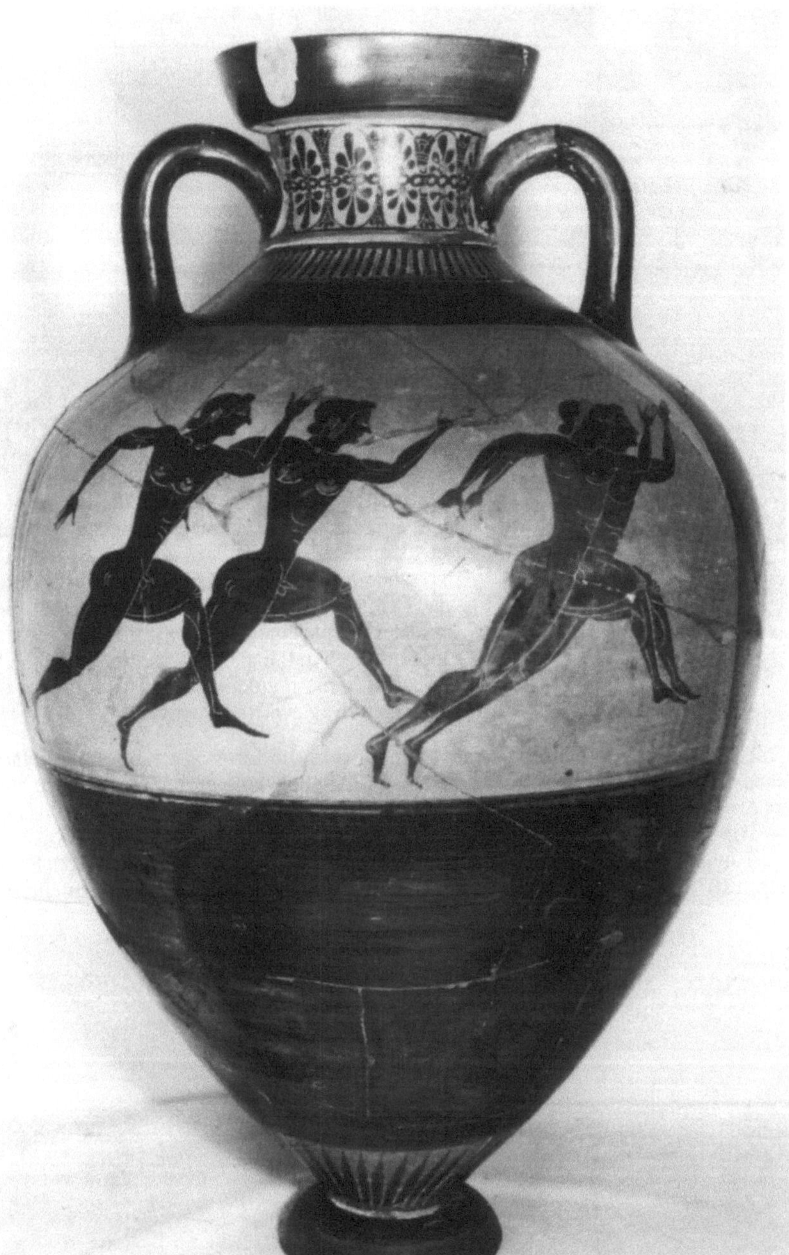

Plate IX. Two Marathon Runners. TWO MARATHON RUNNERS. Breaking of translational symmetry and variation of the momentum, in the Attic Panathenaic amphora from the museum at Vulci (n. 64220)

In this work, the idea of accelerated motion arises from the *dynamics* that go with the *breaking of translational symmetry*: the racer running behind, in the effort to take a *longer stride* than the leading racer, gives a *greater momentum* to his race

We will next illustrate the Stark effect in the hydrogen atom, deferring its technical aspects to Appendix II and Fig. 17, where the results of the Appendix are summarized.

The electron's potential energy in the potential of the proton, has spherical symmetry: whatever the position of the electron may be on a spherical surface centered on the proton, the potential energy of the electron remains the same. We also know that this energy does not change if the electron is overturned with respect to the proton: the parity transformation, $\mathbf{r} \to -\mathbf{r}$, is a symmetry transformation.

Symmetry of inversion, or parity

What happens then when the hydrogen atom is immersed in an electric field E_z, parallel to the z axis, for example? Obviously, a privileged direction and orientation are created.

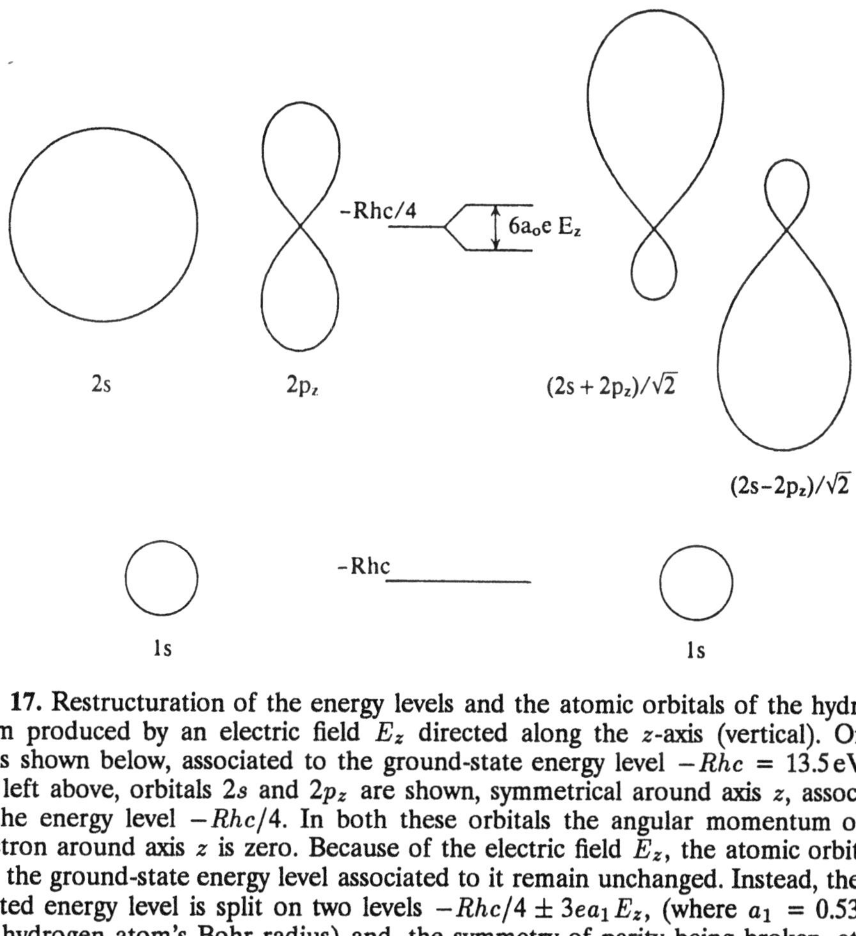

Fig. 17. Restructuration of the energy levels and the atomic orbitals of the hydrogen atom produced by an electric field E_z directed along the z-axis (vertical). Orbital $1s$ is shown below, associated to the ground-state energy level $-Rhc = 13.5\,\text{eV}$. On the left above, orbitals $2s$ and $2p_z$ are shown, symmetrical around axis z, associated to the energy level $-Rhc/4$. In both these orbitals the angular momentum of the electron around axis z is zero. Because of the electric field E_z, the atomic orbital 1s and the ground-state energy level associated to it remain unchanged. Instead, the first excited energy level is split on two levels $-Rhc/4 \pm 3ea_1E_z$, (where $a_1 = 0.53\,\text{A}$ is the hydrogen atom's Bohr radius) and, the symmetry of parity being broken, atomic orbitals $2s$ and $2p_z$ combine to produce hybrids $(2s + 2p_z)/\sqrt{2}$ and $(2s - 2p_z)/\sqrt{2}$, as indicated on the right in the illustration.

To minimize the potential energy the proton, positively charged, tends to follow the electric field, moving north, let's say, while the electron tends to move south, opposite to the direction the field points in. In short, it is as if the electron preferred the hemisphere south of the proton. The very fact that the northern hemisphere can be distinguished from the southern implies however breaking the symmetry of parity. Because of the perturbation provoked by the external electric field, atomic orbitals that were initially symmetric or antisymmetric combine with each other: the hybrid orbitals that result choose the applied field's direction of application as the direction of symmetry (Fig. 17). The original energy level breaks up, and the spectroscopic and dynamic behavior of the atom is altered. New spectrum lines appear: the Stark lines.

Breaking the symmetry of parity and hybridization of atomic orbitals

The Stark effect therefore consists in the restructuring of an atom's energy levels, caused by a static electric field, and in the resulting appearance of new spectrum lines in spectroscopic measurements made on an atom immersed in the aforementioned external electric field.

The restructuring of energy levels produced in an atom by applying a static electric field is combined with a breaking of symmetry.

Seen in this perspective, as we mentioned earlier, even the action of the electric fields, produced by one atom on another as they move closer to each other, which pre-shapes the forming of a molecule, may be interpreted in terms of a partial breaking of the initial spherical symmetry belonging to each of the two atoms taken individually: in fact, the valence electrons of atoms which, drawing near each other contribute to forming a molecule or a crystal, find themselves immersed in a potential which has reduced symmetry compared to the spherical symmetry of the isolated atom.

The breaking of symmetry which results may – at the level of individual atoms – allow atomic orbitals to mix, that is, hybrids to form, the lobes of which point along what proves to be the bonding directions.

It is natural for the formation of atomic orbital hybrids to accompany splitting and/or variations of the energy levels of the valence electrons. The bonding molecular orbitals, in each of which the electrons arrange themselves in pairs, are then obtained by combining the hybrid orbitals of contiguous atoms linearly; the latter must have symmetry characteristics in common with respect to the axis of the bond. An example of hybridization (trigonal) is illustrated, for the benzene molecule, in Fig. 18.

The benzene molecule

To sum up, we've been dealing until now with the effects of a static electric field on the structure of atomic orbitals and their related energy levels. If the perturbing electric field is applied externally, the Stark effect results. If the electric field originates in the atoms' gradually drawing near each other, hybridization occurs, that is, the valence orbitals of each atom are mixed; as they move closer still, another combination takes place. This time the "ingredients" are the hybrid atomic orbitals belonging to contiguous atoms: this produces the molecular orbitals into which the electrons forming the covalent bond flow. Both in the Stark effect and in hybridization phenomena, the breaking of symmetry is a determining factor.

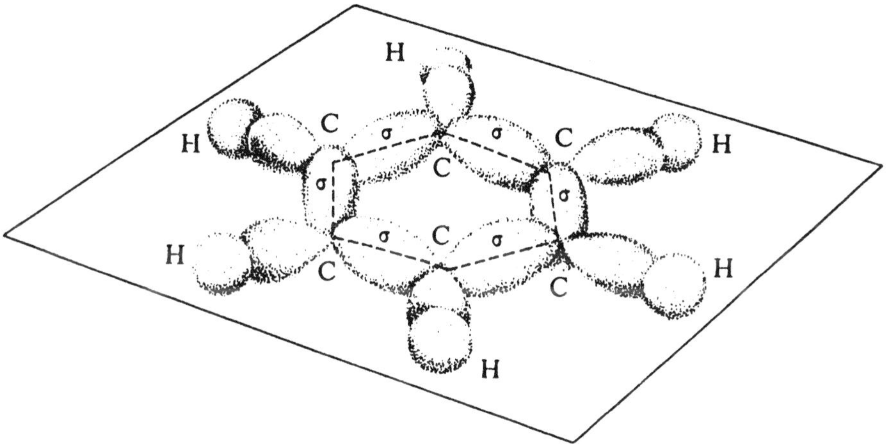

Fig. 18. σ molecular orbitals in benzene. This is a planar molecule, C_6H_6, composed of six carbon atoms and six hydrogen atoms, ring-shaped on the plane (x, y). The atomic orbitals $2s$, $2p_x$ and $2p_y$ of three of the four valence electrons of each carbon atom, in the field having non-spheric symmetry of the two contiguous carbon atoms of the hydrogen atom, combine linearly changing into orbitals of the $2s \pm 2p$ type. Thus they create three hybrid atomic orbitals which point in three directions at 120 degrees (trigonal hybrids, sp^2). Each of these orbitals combines with the atomic orbitals of the three contiguous atoms (two orbitals, also hybridized, of the carbon atoms, and hydrogen orbital 1s with spherical symmetry), creating σ-type molecular orbitals. There are two electrons in each of these orbitals

But a breaking of symmetry in atoms or molecules takes place also when there is no external static electric field, when a spectroscopic observation is made. This is done by analyzing the (visible) radiation absorbed, emitted or diffused inelastically by the atomic systems. An electromagnetic field which varies periodically with the passing of time is associated to this radiation; its frequency, multiplied by Planck's constant, is selected equal to the difference among the energy levels linked by the spectroscopic transition (cfr. Fig. 6). It's as if the electric component of the electromagnetic field, breaking the molecule's symmetry, hooked onto the electron and guided its transition from one energy level to another, according to a resonant mechanism which, for the hydrogen molecule, leads to the charge transfer spectrum (cfr. Chapter Two, 3). Spectroscopic considerations of this type may be referred to one of the two electrons of the hydrogen molecule or to the nitrogen atom in the ammonia molecule, as well as to other centersymmetric molecular systems.

Magenta

We have in mind, for instance, the case of the magenta ion. This is a symmetric ion (in the sense in which, when there is no radiation, the ammonia molecule is centersymmetric) and, as such, has *g*erade (*g*) and *u*ngerade (*u*) energy levels. A transition between the *g* and *u* levels in this ion is possible only as the result of an external action which breaks the system's symmetry. The magenta red radiation has the right frequency, ν_{gu}, for bringing on the transition $g \rightleftarrows u$, activating a resonant exchange in the optical electron between the left radical and the right radical of the above-mentioned ion. The perception of color in the light diffused by this coloring agent is stimulated by precisely this resonant exchange.

Resonant exchange of the electron, and color

In this context it doesn't seem unfounded to liken the spectroscopic mechanisms at the basis of color perception to those at the basis of the dynamic perception of the ambiguity of resonant modular structures.

Graphic condensation

Let us refer to the structure at the bottom of the graphic condensation in Plate IV, for example. The central, inside wall, subjected to visual thinking, is exchanged between the touching cubic moduli alternately, and oscillates. But the electron oscillates too, though with a much more regular periodicity, and piloted by the electromagnetic radiation with frequency ν_{gu}, is exchanged resonantly between the left and right radicals of the magenta ion.

The resonant exchange of the central wall, and the dynamics of ambiguity

Actually, in the above analogy, one must take into account the fact that the perception of an ambiguous figure is realized only after the observer has carried out an elaborate control of the figure. This process consists in collecting the elements mentally which compose the figure and correlating them in coherent schemes, before they slip out of mind. As soon as control of the stimuli exceeds a certain threshold, a dynamic instability is reached which is perceived by the breaking of symmetry and the forming of visual thinking.

A phenomenological model of this perceptive instability is described in the penultimate chapter.

Chapter Four

Entropy and Information

> One should not speak about things which can't be spoken of.
>
> Ludwig Wittgenstein, 1921

In 1948 Shannon ran up against the problem of defining the *informational content of a message*. He discussed it with von Neumann who suggested:

> "You should call it entropy, for two reasons. In the first place, because your uncertainty function was used with that name in statistical mechanics, and therefore it already has a name. In the second place, which is perhaps the more important, no one actually knows what entropy may be, so in any discussion you'll already be a step ahead."

Even in the field of the so-called exact sciences, words that seem to be thrown in just for the fun of it may stimulate impassioned and long-lasting polemics.

A great deal in fact has been written about *entropy* since Clausius, in 1865, introduced this quantity to distinguish reversible or isoentropic thermodynamic transformations from spontaneous ones, and since Boltzmann, in 1872, provided the statistical interpretation of this concept in terms of – number of – possible choices or uncertainty. But we can predict that even more will end up being written by those who, since 1948, have held and will hold that they cannot ignore the relationships between the physical entropy of statistical mechanics and thermodynamics, and the information or informational entropy of the theory of communications. "No one really knows what entropy is", von Neumann said. Actually, it does seem improbable that anyone should learn all there is to know about entropy, or about uncertainty, and that he should therefore be in a position to remove any uncertainty on entropy, thus furnishing complete information on what is by definition uncertain.

Physical entropy and informational entropy

Concerning the possibility of correlating entropy with information, in the literature one finds attitudes ranging from deep-rooted skepticism:

> "my point of view is that the analogies are interesting and worth studying, but they don't go far enough to reveal new possibilities for cross-fertilizing the two disciplines" (Beck, 1976),

to consensus:

what is absolutely necessary – driving the car with care, so as to avoid abrupt braking, as a cyclist who is naturally sparing of his own muscular energy would do. Ten years after the onset of the energy crisis, even the automobile companies were offering their customers car models which allow for reduced consumption. Since 1973, in short, the number of people who confuse wealth and prosperity with the short-lived availability of those who waste non-renewable natural resources, has diminished considerably. There's a certain satisfaction in this.

The energy crisis

If at this point we rub our hands together, we feel a sensation of warmth. The mechanical work developed by the hands, forced to slide against each other, takes on the form of thermal energy and goes on to activate the thermal motion of the atoms, causing an increase in the temperature of the hands. In its turn, the heat development set off by mechanical work is associated to entropy production (in this case, thermal disorder): thus we owe the fact that the mechanical work results in a pleasant sensation of warmth rather than painful skin abrasions to this production of entropy.

Energy dissipation and entropy production

The thermal channel, activated by entropy-producing mechanisms, is actually there, and protects like a safety valve. Fortunately, it is capable, ready and insatiable regarding the dissipation of mechanical work. And heaven help us if this were not so: if mechanical work found no outlet itself in entropy, that is in disorder, every material would be as fragile as glass, and unsuited to the structural functions that mechanical engineers and project designers usually assign them, using them in construction.

There are two sides to every coin, however. Experience suggests that the reverse process – converting thermal energy into mechanical work – may in fact occur, but it requires a thermal machine whose performance is always less than one, even when working under ideal conditions. And we all accept the idea, for example, that obtaining the mechanical energy necessary for starting up the rubbing motion requires more than merely warming one's hands.

Energy dissipation and irreversibility

These observations fall within the framework of the first and second principles of thermodynamics. This section in fact presents a classical formulation of these fundamental principles.

The *first principle of thermodynamics* establishes that energy under any form is conserved:

The first principle of thermodynamics

if a quantity of energy Q in the form of heat is introduced into a system and mechanical work W is carried out on the system, the sum of $Q + W$ constitutes the increase in the system's internal energy ΔU:

$$\Delta U = Q + W.$$

Thermal energy is therefore a form of energy. It is however notoriously less "flexible" than electric energy, and likewise less "noble" than mechanical or gravitational energy. The sea, for example, is a vast reservoir of thermal energy. But this doesn't mean that it will no longer be necessary to burn fuel to keep ships moving.

The *second principle of thermodynamics* provides a criterion for qualifying various forms of energy, or rather, for classifying the ways they may be used:

it is impossible to realize a transformation which has as its *only* result the transformation of heat, drawn from a single source at uniform temperatures, into work.

The second principle of thermodynamics

In this, Lord Kelvin's formulation of the second principle, the transfer of heat from a system at a certain temperature takes on a fundamental role.

Heat may be converted into mechanical energy, as occurred in the steam engine: only partially however, and providing it uses a thermal jump, that is, a source (boiler) which supplies a quantity of heat Q_1 at temperature T_1 and a heat sink (cooler) that absorbs a quantity of heat Q_2 at a lower temperature, T_2.

If the heat transfer in a thermal motor takes place violently, entropy-producing processes are favored and the production efficiency of the mechanical work is relatively low.

Heat transfer may however take place slowly, that is, in a nearly static manner, or reversibly. In the first case, the source and the thermocarrying fluid pass through a succession of states of equilibrium: in order for this to happen, the heat transfer must take place over a period that is relatively long compared to the so-called relaxation times, which are necessary for attaining local equilibrium. In the second case, the total entropy of the fluid-source system remains constant: as the entropy of the source at temperature T_1 decreases, that of the thermocarrying fluid increases accordingly. In both cases, efficiency is relatively high: as we shall see shortly, it is never unity.

This doesn't mean however that a heat transfer is always possible. In fact,

it is impossible to perform a transfer which has as its *only* result a heat transfer from a system at one temperature to another system at a higher temperature.

The second principle of thermodynamics

This is an expression of the second principle, formulated by Clausius, equivalent to the previous statement by Lord Kelvin. This too places limits on the utilizability of thermal energy, and once again the heat transfer and the temperature at which it occurs assume an essential role.

Let us focus our attention on the relationship that exists between heat and temperature.

When a mother places her own upper lip on her child's forehead to see if he has a fever, it may be that the forehead actually "burns" a bit: a certain quantity of heat Q_1 is then removed from the forehead at temperature T_1 and slowly transferred to the lip, which has a temperature T_2, slightly lower than T_1.

Suppose that this transfer takes place in a nearly static way. It is then possible for the ratio $(Q_1/T_1)_{\text{nearly static}}$ to assume the meaning of entropy flow from child to mother.

Entropy and information

In this thermoaffective interaction, the small amount of entropy yielded by the child is gained by the mother. She knows how to utilize it effectively to translate it into the information she is seeking regarding her child's health:

thus it is actually as if the removal of entropy, or uncertainty, from the child produces information regarding his state of health: the child's entropy decreases and information on his condition increases. In any case, information can be obtained only at the cost of a variation in entropy.

The same quantity Q/T, capable of triggering complex and delicate information-producing mechanisms, is present in the expression for the efficiency of a thermal machine devised to produce work.

The ideal efficiency of a thermal machine operating reversibly

In 1824, Carnot evaluated the efficiency of a thermal machine working in a *reversible* cyclic manner, so as to regain, at the end of each cycle, the internal energy characteristic of the initial state. He demonstrated that this efficiency, $\eta_{\max}$, though always less than one, always exceeds that of any other thermal machine working – non-reversibly – in the same temperature interval.

For a fluid which absorbs from the boiler reversibly, at an absolute temperature T_1, a quantity $Q_{1\,\mathrm{rev}}$ and – reversibly – transfers a practically non-recoverable quantity of heat $Q_{2\,\mathrm{rev}}$ to the cooler at temperature T_2, the efficiency of the cycle is defined by the ratio between the work W obtained and the heat coming from the boiler. This is independent of the nature of the thermodynamic fluid and the machine's mechanisms; it depends solely on the temperature of the boiler and the cooler and is equal to:

$$\eta_{\max} = \frac{W}{Q_{1\,\mathrm{rev}}} = \frac{Q_{1\,\mathrm{rev}} - Q_{2\,\mathrm{rev}}}{Q_{1\,\mathrm{rev}}} = \frac{T_1 - T_2}{T_1} = 1 - \frac{T_2}{T_1}.$$

Thus the ideal efficiency is always less than unity.

The real efficiency of a thermal machine operating irreversibly

Actually, every cyclic machine operates non-reversibly, and as such is able to transform into work a quantity of heat $Q_1 - Q_2$ *smaller than* $Q_{1\,\mathrm{rev}} - Q_{2\,\mathrm{rev}}$. It therefore has an efficiency η of less than $\eta_{\max}$:

$$\eta = \frac{Q_1 - Q_2}{Q_1} \le \frac{T_1 - T_2}{T_1} = \eta_{\max}.$$

In a typical, medium-power steam power-station, for example, the thermodynamic fluid (water) is heated to above 500°C (equal to 773 K), and the steam condenses at temperatures no higher than 40°C. Excluding the losses from auxiliary services and from the non-negligeable amount of energy spent in constructing the plant and absorbed by the processing of materials utilized for machinery and civil works, the actual efficiency η is little more than 40%, compared to the ideal $\eta_{\max}$ of $(500 - 40)/773 \simeq 60\%$.

The above expressions do not always help to discourage useless efforts to obtain the impossible – perpetual motion, for example – (on this subject, a paradoxical work by Regge (1981) seems effective). Because of their unlimited applicability, these expressions allow us to avoid calculating special cases in detail. Their very universality however limits their usefulness in the many cases where detail, or the kinetics of the mechanisms, or the nature of the "fluid" and its interactions with the "universe", or the numerous reasons for irreversibility play a fundamental role.

The foregoing expressions can be lumped together in the inequality:

$$\frac{Q_1}{T_1} - \frac{Q_2}{T_2} \leq 0,$$

in which the equal sign refers to an ideally reversible cycle for which the efficiency assumes the maximum value.

Instead of cycling the thermovector between a single boiler and a single cooler, one may refer to a continuous reversible cycle, a portion of which is characterized by the relation $dS = (dQ/T)_{\text{rev}}$.

This ratio is called *entropy increase* of the fluid during that portion of the cycle.

If one applies the latter equation to a continuous reversible cycle, one finds that entropy S of the fluid reassumes the same value it had at the beginning: in fact entropy (like internal energy) is a characteristic function of the state of the fluid.

It can also be demonstrated (see e.g. Fermi, 1936) that if the above-mentioned inequality is applied to an irreversible complete cycle – that is, irreversible in the part leading from state A to state B and reversible returning from state B to state A – and if one works on a thermally isolated or adiabatic system, the relation

$$S_B > S_A$$

Irreversibility and entropy production

holds, where S_A and S_B are entropy values in states A and B respectively: the spontaneous and irreversible transformations which occur in an isolated system, go along with an *entropy production*.

Entropy moves toward a maximum. Once the maximum is reached the system assumes a configuration of equilibrium; it is no longer able to undergo transformations or supply work, unless it is forced by external systems to which it is connected.

As the entropy of an isolated system increases, the availability of energy which may be extracted from the system itself decreases. In this sense, entropy provides a measure of the energy's lack of quality.

Commenting on the relation $S_B > S_A$, Clausius wrote (1865):

> "if we wish to indicate an appropriate name for S, we may say that it is the body's transformation content, in the same sense in which we say that quantity U is the heat and work content of the body itself. Still, since I feel it's better that names for quantities which, like this one, are important for science, be chosen from ancient languages, I suggest using the name entropy of the body to indicate quantity S, from the Greek word ἐντροπή, transformation" (source: Beretta, 1979).

A certain amount of attention is needed to single out the isolated system in which an entropy increase is taking place. The entropy of a cup of hot tea diminishes when it is left on the table and cools down. But this decrease is more than compensated for by the increase of entropy caused in the immediate surroundings by the intake of water vapor and the irreversible transfer of heat to the table by conduction.

The transfer of heat from a hot region to a cooler one in an isolated system is therefore a spontaneous and irreversible process, characterized by an entropy increase in the entire system.

Toward a maximum of entropy while reaching equilibrium

Other typical processes similar to this that we know of are those which spontaneously, with the *increase of entropy* up to a maximum value and the *passing of time*, *irreversibly* lead the milk and coffee to mix in caffe-latte, a wisp of smoke to diffuse in the air and disappear, a swing, after the child has jumped off, to come to a slow stop: these are evolutions of the systems toward equilibrium situations in which transformations no longer take place, entropy reaches a maximum, and "the world silences all things and speaks of them no longer".

2. Statistical Significance of Entropy: Disorder and Disinformation

Entropy and disorder

This section presents the statistical significance of entropy in terms of total amount of disorder of a structure at the microscopic level when the atomic units that form it are given a certain freedom of choice. Entropy in fact represents the number of choices that lead to all the microscopic conformations (configurations) subtending the macroscopic equilibrium structure. In order to analyze the statistical and thermodynamic significance of this magnitude more thoroughly, we will examine two transformations: the order-disorder transition in a binary alloy, and an isentropic, thermoelastic deformation in rubber.

Let us refer to a binary alloy, CuZn, composed of two species of atoms, for example copper and zinc. In this alloy, the atoms are situated on the nodes of a body-centered cubic lattice. Arranged in this way they can create two types of structures: an ideally ordered structure and a disordered structure.

An ordered alloy

In the *ideally ordered* structure (Fig. 19a) the copper atoms, for example, place themselves on the vertices of each cube, and the zinc atoms in the center: in that case, contiguous atoms of both kinds alternate themselves systematically and in order along the diagonals of the cubic cells, according to the sequence

... Cu Zn Cu Zn Cu Zn Cu Zn Cu Zn Cu ...:

no reader would find it difficult to continue this redundant succession of symbols.

A disordered alloy

Instead, *in any one of the disordered structures* (Fig. 19b), in each lattice position, at the vertices or the center of the cubic cells, there are on the average as many atoms of copper as of zinc: but on the average, unlike what occurs for the ordered alloy, there is no correlation between the presence of one kind of atom in a lattice position and the presence of the same or the other kind of atom in contiguous lattice positions. As this correlation decreases, there is a gradual passing from the ideally ordered and redundant

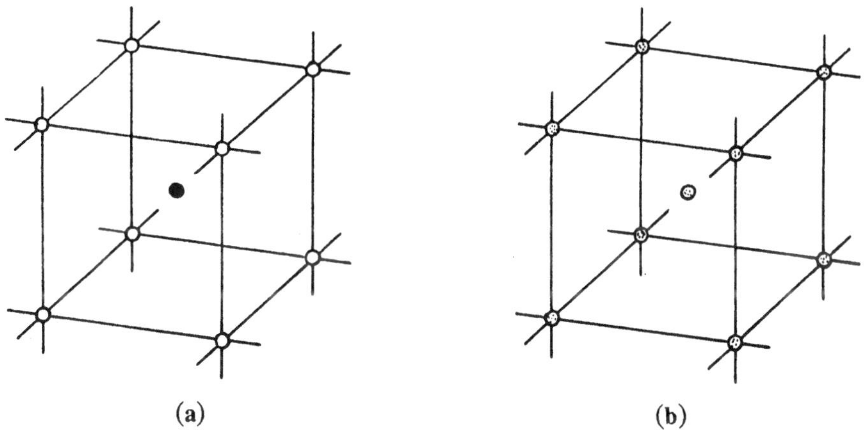

Fig. 19. The elementary cell of an (**a**) ordered, (**b**) disordered copper-zinc alloy sample. In the disordered alloy there is equal probability of finding an atom of copper or zinc in any lattice position

structure to a completely disordered (or uncertain) structure, abounding in configurational entropy (or even, as we shall see, in informational entropy).

The total amount of disorder in the alloy structure increases with the temperature and is controlled by two competing factors, the first fostering order and the second disorder.

The first of these two factors is composed of

Interatomic attractions fostering order

- the relative values of interaction energies among contiguous copper atoms, contiguous zinc atoms and contiguous atoms of copper and zinc. In order for the alloy to form without atoms of the same species segregating into pure copper or pure zinc crystallites, the attractive potential energy of different species of atoms must prevail over the energy (also attractive) among atoms of the same kind. The energy factor, which for the alloy considered here leads contiguous atoms to attract each other particularly if they belong to a different species, is countered by the second factor, that is

Thermal energy fostering disorder

- free choice of the lattice sites prepared to receive the atoms of the alloy. As we shall see, it is as if the atoms were actually able to exercise this choice, liberating themselves from the previously considered energy conditions, the degree of freedom increasing with the temperature of the alloy. Actually, above a certain temperature, called critical temperature, the alloy never has an ordered configuration.

Appendix III contains an application of combinatorial analysis for determining the number of microscopic patterns – configurations – which subtend two extreme structures of the alloy:

- the totally disordered macroscopic equilibrium structure, in which the disorder-causing factor clearly prevails over the order-causing factor: this applies when the equilibrium temperature is higher than the critical temperature, T_c, at which the order $\rightleftarrows$ disorder transition occurs;
- the ideally ordered macroscopic equilibrium structure where, vice versa, the order-causing factor dominates the disorder-causing factor: this applies when the equilibrium temperature is lower than the critical temperature.

The number of alloy configurations is linked to the freedom of choice (Zn or Cu) belonging to each individual crystallographic site inside the alloy itself, and therefore to informational entropy.

Equivalence of physical entropy and informational entropy

On the other hand, specific heat, temperature and physical entropy are bound to each other at equilibrium. By measuring the temperature dependence of the alloy's specific heat, it thus becomes possible both to ascertain the increase of physical entropy during the transformation from the ideally ordered macroscopic configuration to that which is totally disordered, and subsequently, to compare this increase with the increase of informational entropy obtained by combinatory analysis.

The Maxwell's demon

In the light of the solution suggested by L. Brillouin for the paradox of Maxwell's little demon as well (see Appendix III), the result is as follows: if one attributes a content of informational entropy equal to the product of Boltzmann's constant K_B by the natural logarithm of two to each binary choice (that is, to the bit Zn or Cu), the following equivalence relation between physical entropy and informational entropy or uncertainty is valid (L. Brillouin 1956, 1964):

$$1 \text{ information bit} \rightarrow K_B \log_e 2 = 0.97 \cdot 10^{-23}\,\mathrm{J\,K^{-1}\,mole^{-1}}.$$

In this perspective, freedom of choice – and informational entropy with it – increases at the same rate as disorder – and the alloy's physical entropy – increases. Moreover, generally speaking, the acquiring of information – or the removal of uncertainty or informational entropy – in one part of an isolated system is possible only if, at the same time, the overall increase in physical entropy of the entire system is no less than the local decrease in informational entropy.

Entropy, disorder, uncertainty, disinformation

Entropy therefore is a quantity of statistical nature, connected to the concept of disorder as well as to the concept of uncertainty or disinformation.

To the reader who is still unsure regarding the various possible meanings of entropy – disorder, relation between heat exchanged reversibly and temperature, measure of informational content, measure of the degradation of energy or its lack of quality – we would like to suggest an experiment which will allow him to perceive directly, in person, the effect of variations, in contrast with each other, of vibrational entropy, ΔS_v, and configurational entropy, ΔS_c, during the thermoelastic deformation of rubber.

It's a very simple experiment, for which one needs only a rubber band.

Perceiving entropy variations while deforming a rubber band

Let's go find this rubber band then, and place it for a few seconds against the thermometer which is our upper lip, so as to record the reference temperature (room temperature, actually). Then stretch the band abruptly and place it promptly against our labial thermometer again: we will feel that the rubber band is just noticeably warmer. After waiting a few moments, so as to give the band (still stretched, but no longer in contact with the lip) time to return to room temperature, relax the tension abruptly, and lay the band promptly on the lip once more: the sensation of coolness clearly felt at this point is due to the *thermoelastic effect*. This effect, which we have just reproduced, was discovered by Lord Kelvin in 1851. It consists of temperature variations, which originate at the same time as deformations produced in a material, and depend on the amplitude and speed of deformation as well as on the nature of the material itself.

The thermoelastic effect

The reason for this behavior can be explained if we remember that rubber is a polymeric material, composed of a tangled mass of long macromolecules (of polyisoprene, for example) interconnected here and there by transversal bonds. The structure of rubber resembles that of slightly overcooked fusilli, heaped in a plate. When the rubber is stretched, the macromolecules tend to lengthen in the direction in which they are pulled, and take on, like fusilli hanging from a fork, a more ordered configuration than the initial one. As we know, entropy provides a measure of disorder, or the lack of correlation between the moduli forming a structure (for rubber, the macromolecules of polyisoprene). This thermodynamic quantity tends to increase in isolated systems, but actually remains constant during thermodynamic transformations that are adiabatic ($Q = 0$) and reversible, such as the thermoelastic deformation of rubber: these are thermomechanical transformations which occur too rapidly for the material to have time to exchange heat with the environment, but slowly enough to avoid irreversible processes of viscous dissipation of the mechanical energy in thermal energy. In rubber (as in the alloy CuZn), entropy is composed of two parts: a configurational entropy, S_c, associated to structural disorder, and a vibrational entropy, S_v, associated to the disorder of internal atomic movements of thermal agitation. At equal total entropy, during an order-producing isoentropic lengthening in the configuration of the molecules, S_c decreases to the same degree that S_v – together with the temperature – increases. Inversely, during an isoentropic compression, temperature diminishes: exactly as we have just discovered.

Incidentally, the rubber band's behavior is part of a highly varied phenomenology, still little known even among those working in the field, which one is confronted with in the laboratory when, in stressing a structural material mechanically, one doesn't simply measure, as usual, only the deformations, but the concomitant temperature variations are recorded as well. Exploring the thermomechanical behavior of matter and interpreting its causes on the atomic scale in spectroscopic and synergetic terms is a stimulating intellectual

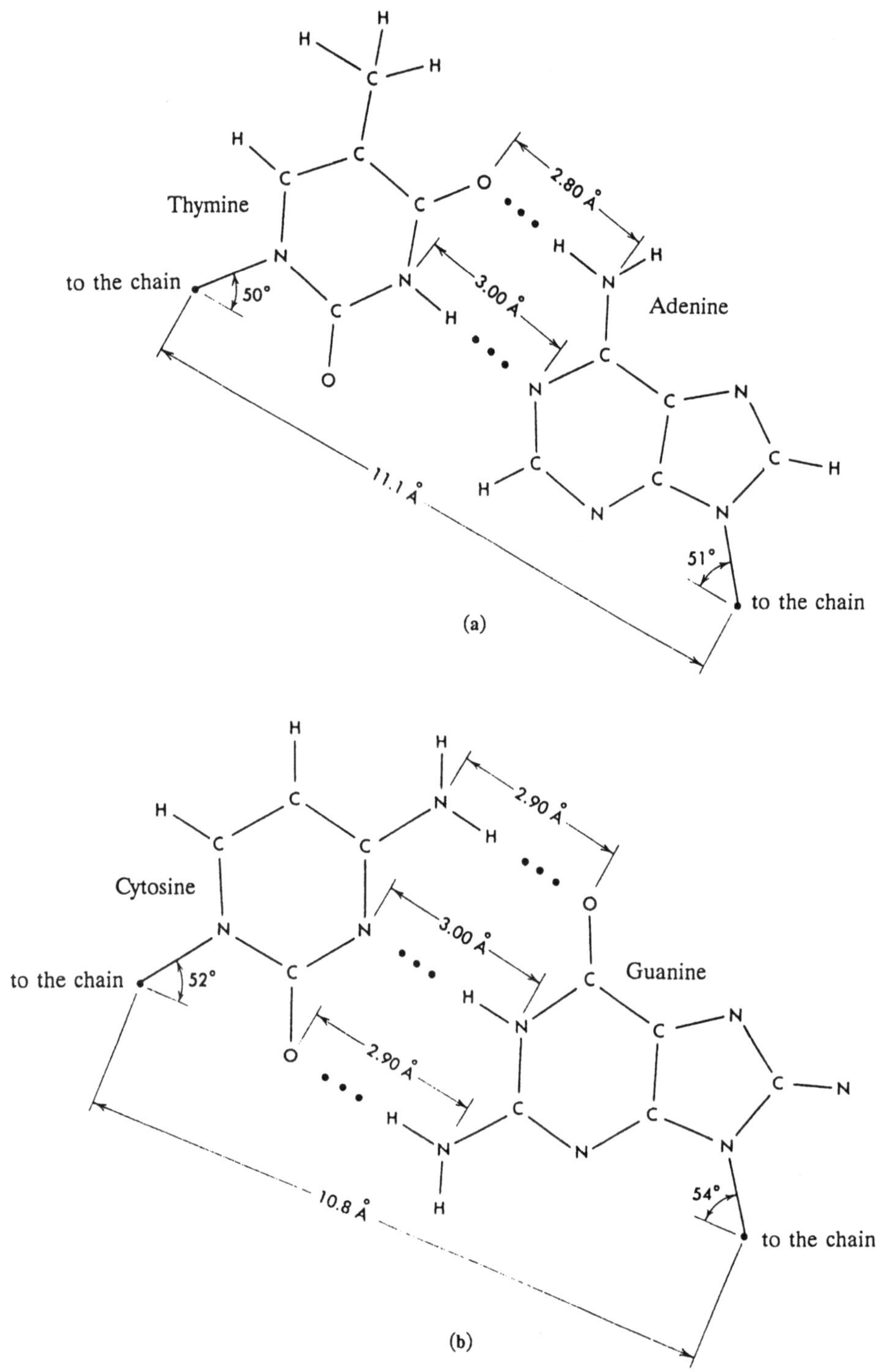

Fig. 21. Pairs of complementary nitrogenous bases of deoxyribonucleic acid (DNA). The adenine (A) and the thymine (T) are linked by two hydrogen bonds, N-H···N and O···H-N. The guanine (G) and the cytosine (C) form a more stable pair, being connected by three hydrogen bonds (see Fig. 33). (From A. Lehninger: Bioenergetics, W. A. Benjamin, Inc., Menlo Park 1971)

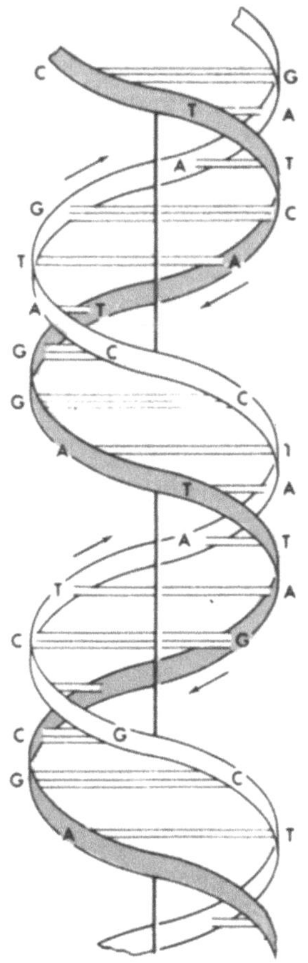

Fig. 22. Sketch of the double helix of DNA. Coupling between neighboring bases, A = T and C ≡ G, belonging to the two chains of DNA is shown. (From A. Lehninger, op. cit.)

of the order of 2 bit/symbol. One may deduce that more or less 2.7 of the 4.7 bits associated a priori to each symbol are redundant.

Contributions to redundancy

This redundancy is partially due to the semantic correlation between the various parts of a piece of writing in which the concepts are expressed with a minimum of coherence, and partially to the rules as a whole which, incorporated at the structural level into the language, represent in the language, as Haken would say, a parameter of order which enslaves human beings: "a language changes only slightly during the course of an individual's life. After his birth he learns a language, that is he becomes its slave, and during his lifetime he contributes to the survival of the this same language."

Leaving aside the cultural contributions to redundancy connected to the significance of the message, which we are not prepared to evaluate quantitatively, we will limit ourselves in the following to classifying the main types of redundancy inherent in the structure of the language. For this purpose, we should correlate the definitions of informational entropy, or the informational content of a message, and of redundancy, with the definition of entropy.

Entropy of a message

As mentioned at the beginning of this chapter, physical entropy (number of binary choices which subtend a macroscopic state of a structure) and informational entropy (number of binary choices which subtend a sequence of letters of the alphabet) have the same formal expression. We have seen that an entropy $K_B \log_e 2$ or (in units $K_B \log_e 2$) $\log_2 2 = 1$ binary choice belongs to each atomic modulus of the structure of the alloy CuZn. Similarly, $\log_2 27 = 4.7$ binary choices belong to each symbol of an alphabet of 27 equiprobable letters (including the blank space).

By examining this page closely one finds that of the almost three thousand symbols that compose it, *a* appears about 170 times, while *q* only appears 16 times. Moreover, while the symbols following *a* may include more or less frequently all the symbols of the alphabet (the fleeting appearance of ah, an improbable digraph, in a piece of writing would immediately lead us to exclude that it be counted as serious literature), one may be sure that only the symbol *u* may follow *q*, as long as we're not talking about the symbol *q* itself, or about Ira*q*, or Sa*qq*ara.

These examples confirm that whenever the symbols are not equiprobable or independent of each other, the number of choices needed a priori to find them decreases: the expression of the informational entropy of a message is less than that, equal to $\log_2 27$ bit/symbol, of a stochastic typescript typed by a monkey.

Thus the symbols of ordinary language are partially redundant.

Redundancy of language

Redundancy is defined in fact as *per cent reduction of the number of choices, starting from the greatest number (of choices), with respect to equiprobable and uncorrelated symbols.*

For example, the parlor game we referred to a little while ago indicates that the redundancy of ordinary language is $(4.7 - 2)/4.7 \simeq 58\%$: a piece of writing from which 58% of the symbols have been cancelled at random, is still at the limits of legibility. Analogously, the informational entropy of a disordered alloy, where the elements that compose it are present in unequal concentrations, is less than that, equal to $\log_2 2$ bit/atom, of a disordered CuZn alloy.

Shannon (1948) and more recently Gatlin (1972) analyzed the nature of redundancy in, respectively, ordinary language and genetic language. They isolated and identified two qualitatively different contributions to redundancy:

a) the contribution stemming from the fact that the alphabetic symbols or the nitrogenous bases in DNA do not recur equiprobably;

b) the contribution associated to privileged correlations of symbols or bases, in digraphs, trigraphs and so on.

To analyze the first category of contributions, let us look again for a moment at a binary alloy where the elements composing it, 1 and 2, are present in average concentrations p_1 and p_2 respectively. As p_1 tends to one (or p_2 to zero, which is the same), the choices become obligatory. The configurational

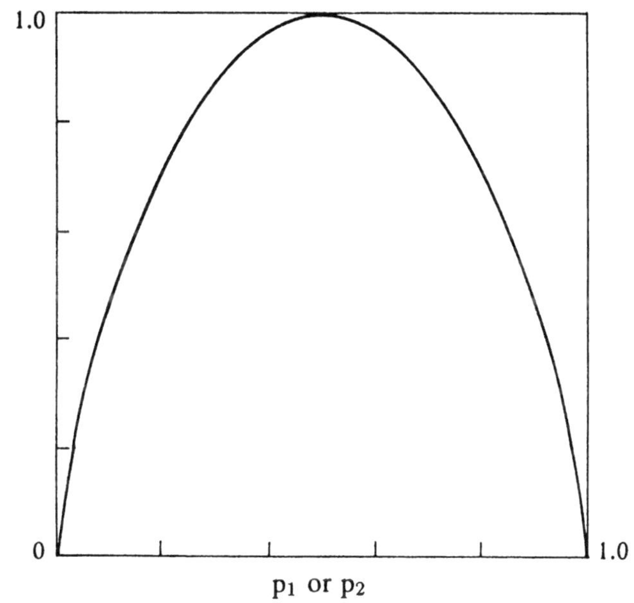

Fig. 23. Dependence of informational entropy per symbol, in a two-letter alphabet, on the a priori probability of the symbol's appearance. The informational entropy is greatest (and equal to 1 bit) if symbols 1 and 2 appear a priori with the same probability, $p_1 = p_2 = 1/2$. In the extreme case in which, for instance, p_1 (or $p_2 = 1 - p_1$) is zero and it is thus certain that the symbol 2 (or 1) will appear, the informational entropy is zero. The above may be applied to the physical entropy of statistical mechanics or thermodynamics – to a binary alloy, for example – as long as entropy is measured in $K_B \log_e 2$ units

entropy of the alloy decreases correspondingly, to the point of vanishing completely when the second element of the alloy is absent. In this case, no bit is needed to decide which atom should occupy each individual site (Fig. 23).

Regarding the second class of contributions, we have already been reminded how for example no bit is necessary to decide which symbol follows the letter q. Generally, there is a correlation among the number of times the alphabetic symbols appear in the different possible digraphs, trigraphs and so on. Linguists have tables on hand which report the average recurrence probabilities of individual symbols, and of correlated symbols in digraphs, trigraphs, etc.

Redundancy in textbooks

In an example of structural analysis of language, Gatlin illustrates contributions a) and b) to redundancy R, drawn from an analysis of the recurrences and correlations of alphabetic symbols in elementary school text books. As the scholastic level of the students for which the textbooks are intended increases, the language used, especially regarding correlation among alphabetic symbols in digraphs, becomes less predictable, and R diminishes (Fig. 24): starting from an initial value of about 0.41, which corresponds to 1.9 bit/symbol, redundancy R reaches a value of about 0.3, which corresponds to 1.4 bit/symbol $(0.3 \cdot 4.7) = 1.4)$. In other words, leaving aside correlations of an order higher than two, on the average about 2.8 bit $(4.7 - 1.9 = 2.8)$ would suffice for pre-

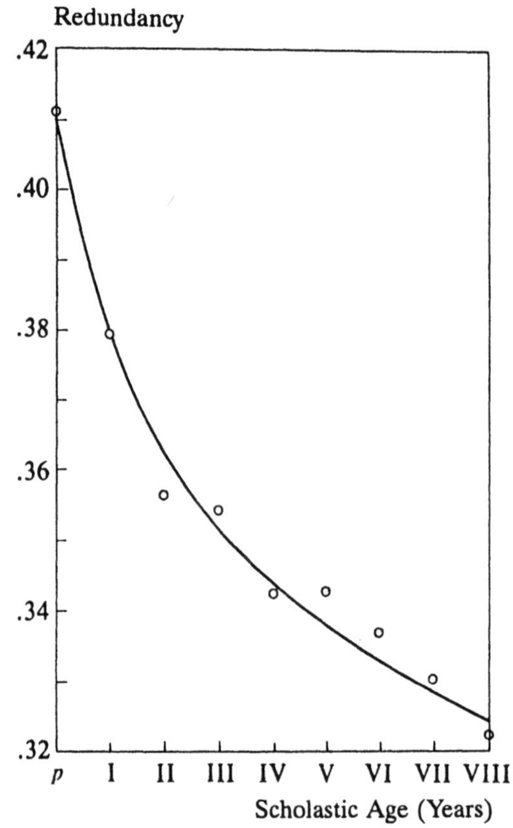

Fig. 24. Redundancy R in elementary school and junior high school textbooks diminishes as the age of the pupils for whom the books are intended increases. (From L. Gatlin: Information Theory and the Living System. Columbia Univ. Press, New York 1972)

dicting an alphabetic symbol in a spelling-book, while approximately 3.3 bit ($4.7 - 1.4 = 3.3$) would be needed for predicting a symbol in a text for more mature children.

Recurrence of alphabetic symbols, correlation among symbols and redundancy

This decrease in R is due not to a variation in the frequency with which the alphabetic symbols recur in the text, but to a progressively reduced correlation among the symbols in the digraphs. In Fig. 25 the decrease of R is broken up into two contributions, D_1 and D_2: quantities which (in unity $\log_2 27 = 4.7$) provide partial redundancies due to the non-equiprobability of the alphabetic symbols and to their correlation at the level of the digraph, respectively. From the illustration one deduces that while the redundancy caused by the different probabilities of recurrence associated to the individual symbols ($D_1/4.7$) remains the same in all textbooks, in those for more mature students the redundancy $D_2/4.7$ diminishes, becoming constant at slightly less than 1 bit/symbol. In interpreting this result, the author observes that in the critical initial phase of language learning, adopting a high redundancy of the D_2 type reduces the possibility of errors to a minimum: it is only when one is immunized against errors that the redundancies may be gradually, and cautiously, reduced.

It isn't possible here to develop these observations to include the contributions to redundancy associated to higher orders of correlation among symbols in trigraphs, tetragrams, etc., or even among complete words. Analyses of this kind are usually conducted with the aid of computers in specialized research centers. They are utilized not only in the literary field, but also in certifying the authenticity of a written work or in identifying an anonymous author.

Following Gatlin's indications, let us now present interesting analogies in the order existing in the sequence of alphabetic symbols in a message and of nitrogenous bases in DNA.

Redundancy of the DNA messages in vertebrates and invertebrates

As happens with the letters of the alphabet, the four bases of DNA or RNA[1] do not recur with the same frequency: that is, they are non-equiprobable, and each of them appears with the frequency typical of its kind. For vertebrates, for example, the frequency of bases C and G falls between 40 and 44%: the fact that $p_{C+G} = 42 \pm 2\%$ differs by 50% implies a type D_1 redundancy, that is, a reduction of the potential variety of the message. Moreover, from a comparison of the field of variability D_1 regarding vertebrates with other species it turns out that the greatest value of D_1 for vertebrates, $D_{1\ \max}$ (vertebrates), is below the greatest value of D_1 for the invertebrates, $D_{1\ \max}$ (invertebrates):

$$D_{1\,\max}\ \text{(vertebrates)} < D_{1\,\max}\ \text{(invertebrates)}.$$

In other words, among all the vertebrates the greatest decrease in variety, $D_{1\,\max}$ (vertebrates), of the nitrogenous bases associated to D_1 reaches a value which, compared to the analogous quantity for all the other species, is minimal. Using a terminology borrowed from the theory of games, the vertebrates attain the minimum of the highest values of D_1, $\min\max D_1$: in the game of the selection of fluctuations which defines the rhythm of evolution, they (the vertebrates) have minimized the handicap that an extreme reduction of the variety of the message represents.

Recalling what we've said regarding the functional importance of type D_2 redundancy in textbooks, it seems natural to make a comparison of D_2 values relating to the genetic message imprinted in the DNA of the various species. D_2 may be analyzed thanks to an experimental method, developed by Nobel prize-winner A. Kornberg, based on the use of phosphorus-32 as a radioactive tracer. Using notation similar to that previously employed to compare the values of D_1 for vertebrates and invertebrates, we found that

$$D_{2\,\min}\ \text{(vertebrates)} > D_{2\,\min}\ \text{(invertebrates)}:$$

[1] RNA is a molecule obtained by transcription; one of the DNA helixes acts as the initial mold. When the RNA is formed, Watson-Crick's rules of coupling are still effective, but the adenine of the DNA, instead of combining with thymine, combines with a new base called uracil, U. Contiguous-base triplets ("codons") of the RNA identify, on the basis of a codification (Gatlin, 1972) which is called genetic code, one of the twenty amino acids which, combined in sequence, form the proteins.

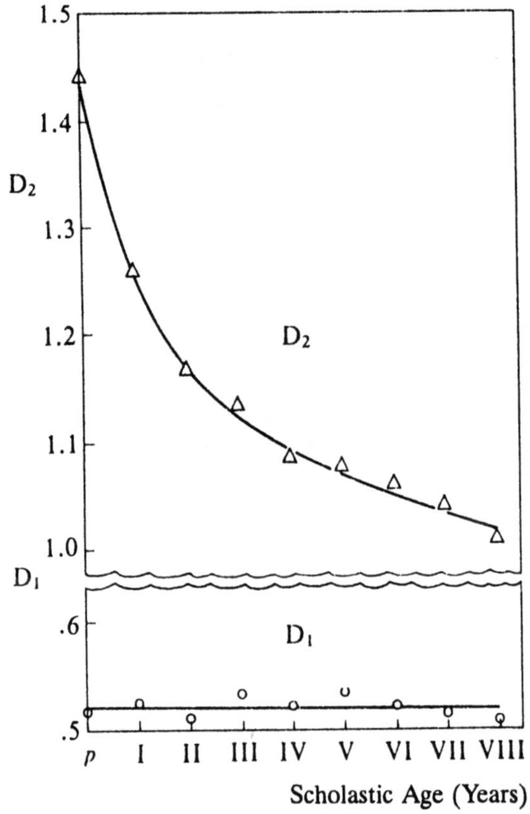

Fig. 25. The redundancy variation which in Fig. 24 is ascribed to the correlation – decreasing with the students' increasing age – between symbols rather than to their non-equiprobability. In fact, while redundancy D_2 due to the correlation among symbols decreases, redundancy D_1 due to their non-equiprobability remains constant. (From L. Gatlin, op. cit.)

that is, during evolution the vertebrates maximized, with respect to other species, the minimum of D_2, reaching $\max\min D_2$: they maximized the advantage represented by a minimum capacity to combat errors of transcription – an advantage since it is accompanied by the greatest variety of digraphs.

Experiments regarding correlations of order higher than two, conducted to identify the sequence of the RNA bases in the tobacco mosaic virus (cfr. Butler and Klug, 1979), confirmed a marked regularity and, locally, a translational symmetry – repetitivity – of the bases themselves. This refers particularly to the sequence (Fig. 26)

Rhythm in the DNA and in the music

...A|GAA|GAA|GUU|GUU|G...

in the curve of the fork in the area where the viral RNA adheres.

This sequence recalls the sequence of the notes in a typical accompaniment to a waltz.

There are about two hundred seventy million possible sequences for the fourteen symbols of a four-letter alphabet ($4^{14} \simeq 268$ million), and conse-

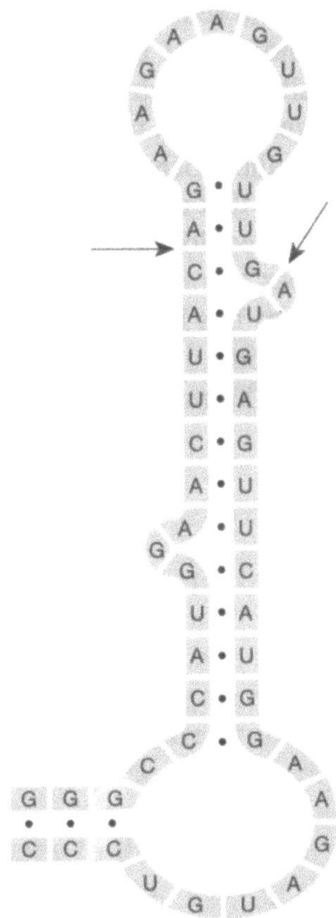

Fig. 26. Sequence of the RNA bases of the tobacco mosaic virus, according to Butler and Klug (1979). The symbols correspond to the initials for adenine, guanine and uracyl. (From J. C. Butler, A. Klug: Le Science, n. 125, January 1979, p. 28)

quently the structural correspondence between one of these sequences and a fragment of the abundant musical production available might not appear purely casual.

With the increase of knowledge regarding the nitrogenous bases of DNA or RNA, we begin to see the possibility of estimating variety and redundancy, symmetry and information in the sequences of genetic language. On the other hand, at present it is already possible (Frölich, 1977) to catalogue, in a kind of bio-dictionary, the *words* which according to the rules of the genetic code may be constructed starting from the codons (that is, starting from the triplets of the nitrogenous bases). Thus the premises exist for discovering the rules of grammar and logic that assign a significance in terms of entity and of biological function to correlated groups of words.

One might ask himself perhaps whether the analogies between genetic language and conventional language, which Gatlin and Frölich presented independently, can be extended to musical language: that is, whether there are

torz, of the "dynamische Resonanzmodularstrukturdoppeldeutigkeitswahrnehmung".

1. Dynamics of the Perception of Ambiguous Structures

Perception of ambiguous figures, selforganization of dissipative structures, spectroscopy of quantum structures

This section rests on the hypothesis according to which the dynamic perception of the ambiguity in resonant modular structures, such as the graphic condensation of two cubes (Plate IV), may in some ways be similar both to the phenomena responsible for the self-organization of dissipative structures carried beyond a threshold of dynamic instability and to the process of spectroscopic observation of quantum structures. Elements regarding this hypothesis are contained in Haken's work (1979) on "formation of structures by dynamic systems and recognition of structures: an attempt at synthesis", and in a monograph by Caglioti (1983) on "symmetry and ambiguity in physics and in visual arts".

At present it seems possible to hazard only a few qualitative observations on the problem of the perception of ambiguous structures. Solving this problem would be an important contribution towards identifying the physico-mathematical bases of the logic of form and graphic language: various cultural milieux, including scientists developing artificial intelligence and robotics, are now calling for development of research in this field (D'Amore, 1979). Leonardo da Vinci was perhaps the pioneer in what he called the "science of painting". Based on his notes "one could construct a theory of pictorial values which is anything but embryonic" (Segre, 1979). A Leonardo who had lived in our times, integrating the modern knowledge of quantum physics and synergetics in his "Treatise on Painting", would quickly have come to fascinating conclusions. And perhaps he would have left his soul in the Mona Lisa, and the Mona Lisa in his soul, just the same.

Let's return to Plate IV. Before going on to analyze the process which leads to perceiving dynamically the binary structures represented here, we should re-read the Inserts and fix our attention on four concepts that will be used during the discussion:

Sensorial stimuli

- the sensorial stimuli, which represent the elements that organize themselves during the perceptive process;

Control parameter

- control parameter σ (the analogue of the temperature reduction in the para-ferromagnetic transformation, or of the inverse temperature gradient in Rayleigh-Bénard's dynamic instability). This measures the control that the observer exercises over the sensorial stimuli caused by the interiorized figure and gathered in his mind, for the purpose of correlating them to each other, synthesizing them in coherent schemes in the memory;

Attention potential

- attention potential V (the analogue of the free energy in a second order phase transformation; this is a function that depends on the order parameter and, parametrically, on the control parameter). It is a dynamic quantity that is characteristic of the interiorized figure, and is determined by the morphological interactions impressed by the designer among the graphic signs forming the structural moduli, by the previous visual experience of the observer and by the value acquired by the control parameter at each moment, during the perceptive process. Fig. 27 shows a typical evolution of the attention potential, for a) relatively low and b) relatively high values of the control parameter;

- order parameter E (the analogue of the magnetization in the ferro-paramagnetic transformation or of the field of velocity and temperature of the fluid strings in Rayleigh-Bénard's dynamic instability). It is visual thinking: faint at the beginning because of low values of the control parameter, it begins to reverberate between "left" and "right" once a certain threshold of the control parameter has been exceeded, and it has plunged into the minimum on the left (or on the right) of the attention potential (Fig. 28).

Order parameter

We should point out here that it isn't possible, at present, to furnish a working definition of σ, V and E, indicating a measuring process for these magnitudes. Thus we cannot deny that, strictly speaking, this problem can be classified among those falling within the sphere of physics only after this stumbling-block has been removed.

The symmetry elements define the essence of a structure

The six binary structures in Plate IV are each composed of a pair of cubic moduli placed side by side. Each structure has an inversion center. This point, exactly halfway between the two cubes of each pair is, although it has not been explicity indicated, an essential point of the structures themselves. One could even declare that, as an element of symmetry, it contributes to defining their essence (Radicati di Brozolo, 1982).

As we shall see, at the moment that the structures formed by the contiguous cubes which then "condense" graphically are dynamically perceived (we are referring to the last three structures at the bottom), the inversion symmetry is broken: the function as center of symmetry represented by the central point of the figure thus loses its significance. But the beauty of each symmetry can be grasped only at the moment it is broken: true beauty is a deliberate, partial breaking of symmetry.

Symmetry breaking: a source of beauty

Let's take things in order.

We will touch first of all on the structures formed by the separated cubic moduli, then on the structure, formed by the two cubes placed in incipient contact, and finally on the "condensed" structure composed by the two cubes having the inside wall of the structure in common.

Using a language borrowed for convenience from quantum mechanics, we shall refer without distinction to the cubic moduli as hydrogen atoms, to their black walls as atomic electrons, and to the "condensed" structure as a hydrogen molecule.

Aggregation of two hydrogen atoms in a molecule

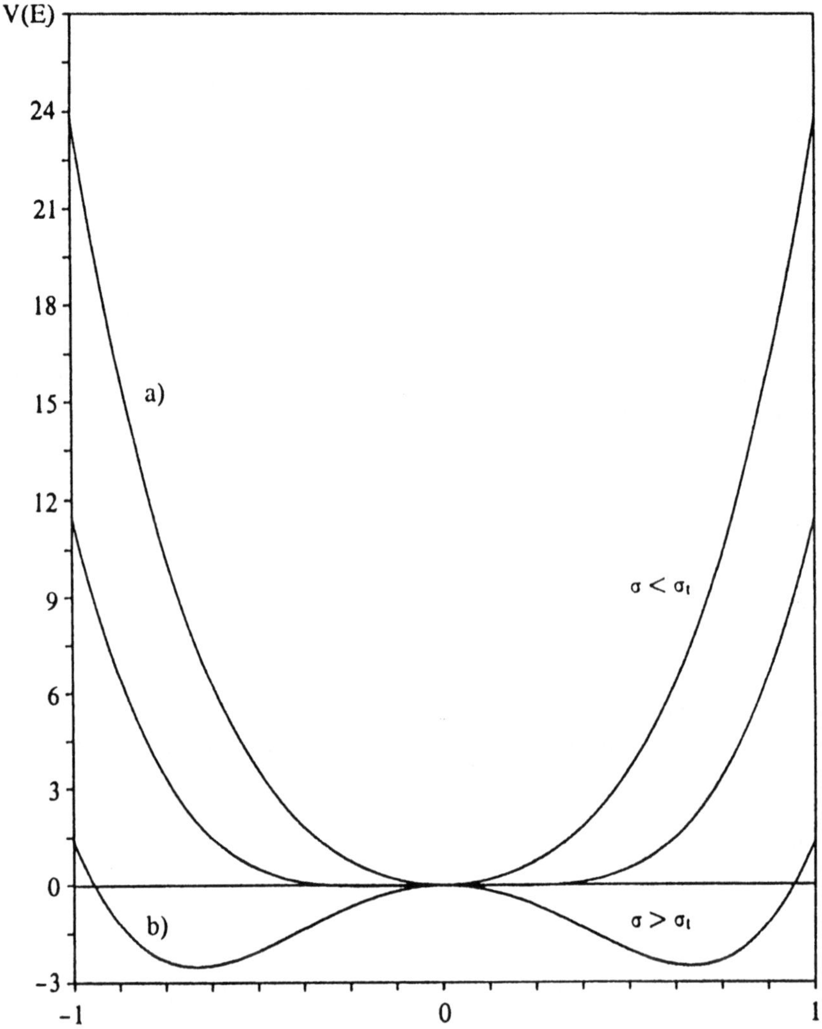

Fig. 27. The attention potential $V(E)$ that an ambiguous figure suggests to the observer's mind depends on the control parameter σ. The cases may be two: a) the control σ that the observer exercises over the sensorial stimuli, trying to store them in mind in the form of coherent schemes, is modest ($\sigma < \sigma_t$). In this case the stimuli tend to slip the mind, the attention potential offers no attractors, and the mind has no way of elaborating visual thinking, whose amplitude E in fact remains null. b) the control σ reaches and exceeds a critical threshold value, σ_t ($\sigma \geq \sigma_t$). The sensorial stimuli organize themselves in coherent schemes g and u, the attention potential's minimum becomes increasingly flatter and wider until it splits into two minimums – attractors – one to the left and the other to the right (cfr. also Fig. 13, c), while visual thinking undergoes widening fluctuations until it "falls", by chance, into one of the attention potential's two attractors – to the left or to the right. At the same time visual thinking mixes u and g breaking the symmetry, and takes the form of the scheme $\psi_+(t=0) = (g+u)_0/\sqrt{2}$ or $\psi_-(t=0) = (g-u)_0/\sqrt{2}$. Once it has fallen into the attractor on the left or the right, visual thinking hooks onto the wall shared by the two cubic moduli just as a resonant electromagnetic field that promotes the charge transfer spectrum would hook onto the nitrogen of the ammonia molecule or one of the two electrons of the hydrogen molecule, transferring it from left to right, then to the left again and so on.

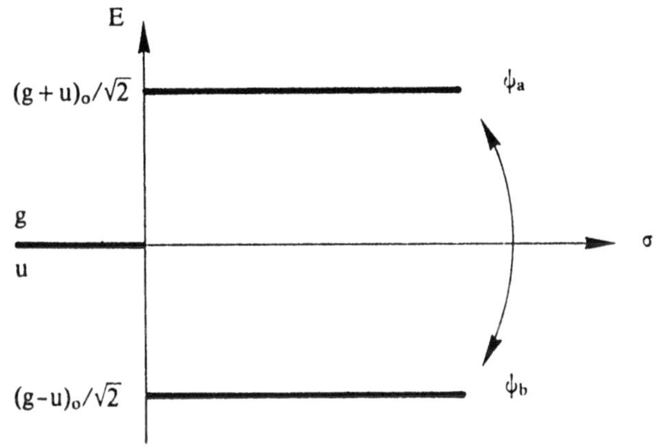

Fig. 28. When the control parameter σ is altered, the figure which in itself is centersymmetric, characterized by a *g*erade g or an *u*ngerade u state, organizes itself before the observer's eyes in an initial state $(g + u)_0/\sqrt{2}$ or $(g - u)_0/\sqrt{2}$; correspondingly, the wall common to both cubic moduli moves to the left or to the right respectively. Having thus reached one of the two bistable focuses in a state of the ψ_a or ψ_b type, visual thinking E next moves the wall into the other focus and then back to the first again, in an alternation of inversions of perspective

Separate Cubes

Graphic condensation of two cubic moduli in an ambiguous structure

The four upper structures in the plate have an inversion center of symmetry. The "atomic" moduli which gradually draw nearer to each other in the four structures, begin to enter into contact only in the fourth structure and do not appear to be connected with each other except for the fact that they form, as a pair, a centersymmetric structure: the "electron" belonging to each of these moduli "is" the side surface (in black) of each cube, and is localized on it. Still using quantum terms, to each of these cubic moduli a stationary state may be associated, ψ_a and ψ_b respectively for the wall belonging to cube a on the left and cube b on the right, that is, for the electron of the atom on the left and of the atom on the right.

The centersymmetric stationary states

In other words, returning to the image suggested by Mott (p. 20), we may associate a pendulum to each of the two moduli. The oscillation amplitude of pendulum a corresponds to stationary state ψ_a of the electron, and the oscillation amplitude of pendulum b corresponds to stationary state ψ_b. Both of these amplitudes are stationary, in the sense that they remain constant in time, without interfering with each other.

Incipient Contact

The fifth structure from the top also has an inversion center. In this structure however the two cubes are graphically correlated, since their walls, representing the "electrons", partially overlap. The "electronic" states ψ_a and ψ_b, recognized by the mind independently of each other when perceiving the separate cubes, are again brought to the attention of the observer, who is stimulated to correlate them in this new structure.

The elementary processes through which this correlation between ψ_a and ψ_b is realized are developed according to non-linear, complex kinetics and will be considered further on. For this intermediary structure, such a correlation seems satisfactorily described by following Mott's suggestion again, that is, through the same formalism used to characterize, according to quantum mechanics, possible state functions of the bonding electron of the hydrogen molecular-ion or of the nitrogen in the ammonia molecule. Actually, it's as if the observer, to whom the figure still looks static and centersymmetric, tended to attribute to the figure itself one *or* the other of the two stationary states, ψ_g or ψ_u. For both these states the inner side wall always belongs half to the modulus on the left and half to the modulus on the right.

The central point of the structure: its belonging and not belonging to the two moduli

But there are two ways in which the inner wall may belong equally, as is fair, to both cubes. In state ψ_g the central point of the wall belongs to both moduli. In state ψ_u, this point does not belong to either modulus.

Perception of this figure does not end at this point: it leaves one generally dissatisfied if, more or less knowingly, he is in fact facing a choice (ψ_g or ψ_u?) both in the present critical stage of the graphic condensation and, more urgently, in the final stage examined below.

Contiguous Cubes

This structure, like those considered previously, has an inversion center.

From flat to spatial, through symmetry breaking

To an observer who, knowing nothing of how the graphic condensation has been developed, observes this structure for the first time, it initially appears to be a bidimensional design possessing an inversion center. But what happens if the control the observer exercises over the sensorial stimuli suggested by the figure goes beyond a certain threshold? The interiorized figure reorganizes itself, presenting an entirely new field for the attention to linger upon. And suddenly it seems to come to life, practically leaping off the page. Disturbing a commensurate equilibrium among the structural moduli, the observer, in the moment of transition produced by the breaking of symmetry and equilibrium, goes on to remove the uncertainty inherent in the composition imagined on the surface of the paper, and he feels induced by the morphological interactions imprinted by the designer to produce information: information, for example, which consists in affirming that the inside wall belongs to the cube on the right.

To the question whether, in the critical moment of transition, the structure remains the centersymmetric, bidimensional design or is the tridimensional structure with a well-formed cube on the right, the answer would be: both. *Tertium non datur*? In that moment the two aspects of that single ambiguous reality, the interiorized, gestaltic figure, are mutually exclusive, and yet both co-exist in it: in the originally symmetric figure the symmetry of the structure breaks, and information is created.

The fascination of the ambiguity

The fascination of the ambiguity of this structure does not stop at this point. The chaotic whirling of the will could have begun by assigning the inside wall to the cube on the left instead of to the one on the right. At any rate, whatever the initial option may have been – right or left – the observer cannot help perceiving a constant alternation of inversions of perspective. And once again, to the question whether the inner wall belongs to the cube on the right or the cube on the left, the answer would be: to both. *Tertium datur*?

Let us examine the dynamic evolution of the perceptive process of this "condensed structure" a bit further.

At first the structure appears to be centersymmetric: but the control that the observer exercises over the interiorized figure, although modest, is alert. The observer explores the figure, stopping now on this and now on that graphic element. His control tends to correlate all these elements, recording partial syntheses of them in the memory. The tendency – assisted by control – to arrive at an overall synthetic vision of the correlated elements, is countered by an equally instinctive tendency of the elements themselves to escape from the coherent and ordered schemes in which they are mentally imprinted in the memory.

A symmetry breaking followed by rhythmic inversions of perspective

When, in the competition thus established between these two tendencies, control finally wins out, a new, completely unexplored field offers itself to the observer's attention. He perceives, often in a manifestly traumatic manner, first a breaking of the pre-existing symmetry, then a rhythmic alternating of inversions of perspective.

Let's take another look at these two phases of the perceptive process, beginning with the breaking of symmetry.

As mentioned, the initially disordered sensorial stimuli gather into ordered schemes when under the influence of the control parameter. As soon as the control parameter reaches a critical value, all the stimuli may be considered as grouped together, correlated among themselves in schemes or states that we could call ψ_g and ψ_u (Fig. 11). At the critical transition point disorder $\rightarrow$ order these schemes, while separated by a difference in attention level $E_u - E_g$ (Fig. 10), offer themselves jointly to that which, along with Arnheim, we might call *visual thinking*. At this point visual thinking, E, is able to break the parity symmetry of the structure described by the stationary and centersymmetric states ψ_g and ψ_u; visual thinking, playing the role of order parameter, mixes these states equitably, and re-organizes the structure itself dynamically causing it to fall at first into one of the two states equally probable a priori (bistable focus in Fig. 28): the state $\psi_+(t)$

$$(\psi_g + \psi_u)/\sqrt{2},$$

coinciding, at initial time $t = 0$, with ψ_a, or the state $\psi_-(t)$,

$$(\psi_g - \psi_u)/\sqrt{2},$$

coinciding, at initial time $t = 0$, with ψ_b.

In correspondence to these states in fact the attention level falls to a minimum value. At the same time, the wall shared by the two cubic moduli will originally seem to belong to modulus a on the left or modulus b on the right.

From disordered sensorial stimuli to ordered visual thinking

At this point however the interiorized structure is no longer a jumble of disordered and incoherent sensorial stimuli: it becomes a coherent whole, so far enslaved by that sole degree of freedom which is the order parameter formed by visual thinking, as to identify itself with visual thinking. The latter emerges victorious from the competition, guided by the control parameter, which sees the structure's innumerable degrees of freedom being disputed. And it alone suffices then to describe the behavior of that coherent unicum, that great molecule to which the ambiguous structure is also dynamically comparable (cfr. Figs. 13 and 27).

Inversions of perspective

We can now go on to interpret the second phase of the perceptive process: the phase which, starting from the critical moment $t = 0$ of the breaking of symmetry in favor of the (wall assigned to the cube on the) left or the right, consists of inversions of perspective alternating left $\rightarrow$ right $\rightarrow$ left $\rightarrow \ldots$.

We know that state $(\psi_g + \psi_u)/\sqrt{2}$, with whose value starting from time $t = 0$ visual thinking is definitively identified, is not centersymmetric or stationary any longer. For this state, parity is no longer a constant of motion; its amplitude, like that of the resonant oscillation in the double, paired pendulum, passes from left to right and then to left again.

With due caution we could also refer to the spectroscopic analogy of the charge transfer spectrum, that is to the amplitude of the field of resonant, electromagnetic radiation which hooks the electron of a hydrogen molecule: the exchange of the electron between the two protons, fostered by the radiation, takes place with a frequency equal, in Planck's constant unit h, to twice the exchange integral $(-2K/h)$. In state $\psi_+(t)$ the electron (the shared wall) hooked by the electromagnetic field (visual thinking) apparently belongs at time $t = 0$ to proton (modulus) a, at time $t = -h/(4K)$ to proton (modulus) b, at time $t = -h/(2K)$ to proton (modulus) a again, and so on: it resonates between states ψ_a and ψ_b. In state $\psi_-(t)$, corresponding to the other bistable focus in Fig. 28, the electron (the shared wall) seems to belong at time $t = 0$ to proton (modulus) b instead of to proton (modulus) a, at time $t = -h/(4K)$ to proton (modulus) a, etc., producing in this case as well an alternation of inversions of perspective, out of phase with the former one but practically identical to it.

It goes without saying that the spectroscopic analogy obviously cannot be pushed to the point of introducing Planck's constant into problems of perception. At most one could suggest that:

a) a stationary attention level, because of its very stationariness, does not lend itself to being measured;

b) a quantitative measurement of the separation $E_u - E_g$ between two stationary attention levels can be supplied by the average inversion frequency of the percepts, $\omega \propto E_u - E_g$.

Frequency distribution of the inversions of perspective

Actually, the oscillation of the inside wall between the two moduli follows a temporal progression that proves to be only approximatively periodic: it is characterized by a succession of jumps from one percept to the other, which occur with a distribution of frequencies (Borsellino, De Marco, Allazetta, Rinesi, Bartolini, 1972; and Borsellino, Carlini, De Marco, Penengo, Trabucco, Riani, Tuccio, 1977). This succession can be only roughly described by a "decision equation" – essentially non-linear – such as that utilized by Caglioti and Caianiello (1978). In other words, the distribution of frequencies with which visual thinking, feeding on itself, guides the rhythmic alternating of inversions of perspective of the inner wall between the two cubic moduli, apparently depends not only on the morphological interactions among the elements composing the interiorized figure, which the designer imprints in the figure, but also on the intensity that the observer, aided by his previously accumulated visual experience, is able neurophysiologically to give his own visual thinking.

Presumably, the values of average frequency ω with which the percepts alternate are determined partially by the velocity with which our sight scans the individual structural elements of the figure, partially by the kinetics with which, at the neurophysiological level, the latter are collected and then sorted out into the coherent schemes corresponding to the two percepts, or tend to escape, and partially by the morphological interactions that the designer imprints on the figure. That the latter combine in determining ω is proven empirically by comparing, for instance, the rhythms with which the conflicting percepts alternate in the "plastic permutation in the field" in Plate I, and in the structures in the final stages of the "graphic condensation".

Unbalanced persistence times of perspective inversions for an asymmetric ambiguous figure

The two conflicting percepts in Plate I tend to alternate, with a highly unbalanced relationship between their persistence times. That is, it's as if the attention potential, once the threshold of the control parameter of the stimuli has been passed, presented two asymmetric minima, one deeper than the other; there are two attention levels within these minima. The lower attention level corresponds to the persisting bidimensional state, the upper attention level corresponds to the fleeting photophania.

The two conflicting percepts (left, right) in each of the two final stages of the graphic condensation in Plate IV alternate with a regular cadence instead, ideally metronomic. Still, the commutation frequency of the wall shared by the two *partially touching* cubic moduli is visibly lower than the frequency pertaining to the reversal of the *fully condensed* graphic structure.

Inversions of perspective and charge transfer spectrum

It is interesting to note that a similar relationship would exist among the frequencies of the charge transfer spectrum of the hydrogen molecule, if it were possible to set a greater value than that of equilibrium for the inter-protonic distance, in relation to the completed atomic condensation. In the homonuclear biatomic molecules the difference $E_u - E_g$, which measures the separation between energy levels associated to the bonding and antibonding molecular orbitals, depends in fact on the nucleus-nucleus distance R according to a trend that may be qualitatively deduced – for the hydrogen molecular-ion, for example – from Fig. 10. This difference is proportional to the exchange frequency and, as this figure demonstrates, decreases as R increases.

The situation is not fundamentally different from that regarding the exchange frequency in a molecule such as the ammonia molecule: this frequency is an increasing function of the resonant electrical field (the analogue of visual thinking) (cfr. Feynman, 1965, ch. 9 – p. 8). The periodicity of the inversions of perspective which occur during the process of perceiving an ambiguous figure actually encouraged us to compare this process to that of the spectroscopic observaton of a single molecule: however, as we said, this periodicity is characterized by a broad distribution of the percepts' persistance times. The width of the distribution and the average persistance time vary from one individual to the other.

Processes developing during the perception of an ambiguous figure

In short, it seems possible to affirm that the perception of a centersymmetric ambiguous figure is realized through a controlled succession of elementary processes characterized by non-linear kinetics. This is preceded by the organization of the many stimuli suggested, in disorderly fashion, by the interiorized figure. The result of this task is to break the symmetry of the figure, to reorganize the field of attention potential that the figure offers the observer's mind at the moment of bifurcation, and to enable visual thinking to emerge as the order parameter. The latter falls at first into one of the attention potential minima, attributing the common wall to the cube on the left (ψ_a) or to the cube on the right (ψ_b). Subsequently, in linking the stationary centersymmetric states ψ_g and ψ_u, it resonates between the two minima a and b of the attention potential with an average exchange frequency ω defined by the separation $E_u - E_g$ between the attention levels associated to the above-mentioned stationary states. Contemporaneously with visual thinking and interacting with it, the wall shared by the two cubic moduli alternates rhythmically between the moduli themselves.

The dynamics of ambiguity

In this perspective the dynamics of the perception of ambiguity could end up as a kind of resonant coupling between figure and observer. In other words, it's as if the latter's visual thinking included a band of frequencies that is a priori quite broad. Once symmetry is broken, the dynamics of perception and the perception of the dynamics of ambiguity would consist in the observer's becoming aware of having syntonized his thinking on the frequencies in the neighborhood of the average value ω suggested by the morphological interactions among the elements composing the figure.

Finally, let us note that at the moment symmetry is broken, immediately before the inversions of perspective begin to alternate, it's as if the uncertainty inherent in the centersymmetric figure were removed and an information bit were produced (common wall belonging at the moment $t = 0$ to modulus a or modulus b).

The ambiguous structures we have utilized in order to grasp, at the physiological level, the nature of the dynamic instabilities in self-organizing structures, are in general centersymmetric. In the next section, a generalization to more complex structures of the results obtained so far is attempted.

Pattern recognition and perception

After this book originally appeared in 1983, pattern recognition and perception became important branches of computer science and artificial intelligence.

Evolution of the interiorized figure

The process of perception can be now described e.g. in terms of a transient of the evolution of the interiorized figure. This evolution is regulated by a potential which depends both on the control exerted by the mind on the sensory stimuli (*control parameter*) and, nonlinearly, on the components of the current interiorized figure (on the adjoints of archetype vectors acting as attractors). In particular the oscillations in the perception of an ambiguous figure can be modelled using time dependent control parameters in order to simulate neuron saturation of attention. (T. Ditzinger, and H. Haken, *Oscillations in the Perception of Ambiguous Patterns, A Model Based on Synergetics*, Biol. Cybern. **61**, 279–287 (1989)).

Archetypes

Neural networks

Alternately, the process of perception can be envisaged also in terms of a transient of the evolution of the activation levels of the processing units of a neural network implementing an associative memory. (J. J. Hopfield, *Neural networks and physical systems with emergent collective computational abilities*, Proc. Natl. Acad. Sci. USA, **79**, 2554 (1982), Biophysics).

An updated analysis of these problems has appeared recently in the Springer Series in Synergetics: H. Haken and M. Stadler (eds.) *Synergetics of Cognition*, Proceedings of the International Symposium at Schloss Elmau, Bavaria, June 4–8, 1989, Springer, Berlin, 1990.

2. Symmetry as Persistence of Significance as the Result of Structural Transformations, and the Role of Symmetry Breaking in Poetry, Music and the Visual Arts

A symmetry transformation as a transformation leaving the energy invariant

The definition of symmetry operation in terms of a transformation Ω which leaves the energy operator H of a quantum structure unchanged rests on a comparison, unproductive though it may be, between Ω and H.

The problem arises: is it possible to establish a similar criterion of comparison for the non-quantum structures, for example geometric structures or macroscopic systems removed from thermodynamic equilibrium, or even for schemes or concepts into which the perceptive process and thought are divided (cfr. Fig. 1)?

Important publications, such as the Acts of the symposium on "Symmetries", organized in 1969 by the Accademia Nazionale dei Lincei, and the volume on "Symmetry" (1973), where notes from the international convention organized by E. Agazzi at Venice in 1970 are compiled, bear witness to the need to explain the significance and the role of symmetry and the breaking of symmetry in complex systems. More recently, P. Morrison dedicated a monograph in the volume *Aesthetics in Science* to these subjects, edited by J. Wechsler and published by MIT (1978).

In the following, integrating the contents of the preceding section with an analysis of these works and with some examples in part drawn from them, we would like to indicate a possible line of thought.

Symmetry in poetic structures

First we will try to formulate a criterion for recognizing symmetry and identifying its characteristics, in poetic structures. We will then reinvoke the classic definition of symmetry for gestaltic structures. Finally, we will point out the importance, aesthetic as well, of the breaking of symmetry in poetry, in figurative arts and in music.

As J. C. Helmcke (1973) suggests, a kind of parity symmetry can be instinctively recognized in the opening of a famous sonnet by Petrarca:

La vita fugge | e non | s'arresta un'ora:
(Life flies by | and does not | stop an hour)

and yet it seems hazardous to try to associate an energy operator to this hendecasyllable.

So precisely what does the verse's symmetry consist in?

To answer this we must unfortunately carry out a sort of vivisection of this charming structure. Before proceeding further, let us return for a moment to the example of axial symmetry of the glass. The glass' axis is a symmetry axis inasmuch as if we rotate the glass around it we have no way of knowing that we have made the rotation: it's impossible in fact to measure an angle of rotation around the axis, since one cannot establish an absolute reference for the angles, by making a notch for instance, without changing the structure of the glass. But it is precisely the impossibility of setting a reference for the angles that implies the invariance of the glass' potential energy resulting from a rigid rotation: in its turn, this invariance involves the cancelling of the external forces associated to this rotation and thus, finally, the conservation of the component of the angular moment along the symmetry axis. In short, the unproductive confrontation (that is, the cancelling of the commutator, Fig. 19) between the transformation-rotation and the (potential) energy of the glass leads to identifying what remains constant in the transformation: in this case, it is the component of the angular momentum around the axis of the glass that remains the same, and which then generates the transformation rotation around the axis itself. The latter is a symmetry transformation for the glass.

Parity transformations leaving the significance unchanged

At this point we may ask ourselves what would remain unchanged if, treating Petrarca's line of verse the same way as a glass or goblet, we made parity transformations around its central *and does not*. We would have, for example:

Un'ora fugge *e non* s'arresta la vita
(An hour flies by *and* life *does not* stop)
La vita s'arresta *e non* fugge un'ora
(Life stops *and* an hour *does not* fly by)
Un'ora s'arresta *e non* fugge la vita
(An hour stops *and* life *does not* fly by)

Only the second of the three "verses" obtained as a result of parity transformations made on the original line retains a hendecasyllabic rhythm. We

Transformations leaving the significance unchanged, as symmetry transformations

will discuss the problem of rhythm in other examples, since its importance here is marginal. All three of the above "verses" – this is important – retain a capacity to inform, a significance. A significance which, by the way, appears substantially not unlike that of the original verse. In other words, the comparison between the results of parity transformations of the verse around its central *and does not* and the verse's essential property, its significance, which – we should say – contains all the physics of it, is unproductive.

Energy and significance

Such a confrontation is really just as unproductive as the comparison that is made in a quantum structure between any symmetry transformation in the system and its energy, which – we could say – contains all the system's significance.

Symmetry, or invariance of the significance versus (mental) transformations, as a source of beauty

It is likely that, upon reading a line of verse or a poem, perhaps unconsciously, the mind makes structural transformations, Ω, such as those presented above. The beauty of the verse we were working with may derive in part from the feeling of security produced by the fact that in spite of the change caused by these transformations, a meaning of some sort remains; in part too from the awareness, subliminally acquired, that among the various possible lines obtained in this way, the original one has the most fascinating meaning of all.

I was saying that the rhythm, in the line of verse by Petrarca analyzed above, is of secondary importance. It might be interesting therefore to analyze other poetic structures in which, vice versa, rhythm rises almost to the role of significance.

Virgil, with his pentadactylic hexameters, gives us this possibility. These verses, metrically speaking, have translational symmetry. This symmetry evokes a constant momentum, which being constant, generates translational symmetry and onomatopeically manifests itself in the rhythm. (See Insert at the end of the chapter).

These examples give an idea of the problems encountered when one wishes to extend the definition of symmetry to systems (including macroscopic systems) that are more complex than the elementary physical-chemical structures characterized by a time-independent energy operator, or than simple geometric structures.

The symmetry in architectonic and geometric structures

No difficulties arise for architectonic or geometric structures. We need only to recall the definition suggested by Weyl (1952): a structure is symmetric when, subjected to a certain transformation (for example, when it is reflected on a surface or rotated around an axis) it returns into itself. For this class of structures as well, those configurations are symmetric whose elements retain the same structural relationships as a result of certain (geometric) transformations made in the structure itself.

In the field of perception, those (macroscopic) systems are interesting which, when interiorized, provoke sensorial stimuli which organize themselves in the mind, producing thought. For such systems it appears possible to pinpoint, in the *significance*, the "observable" with which an Ω transformation of the linguistic or conceptual kind must be compared in order to establish whether it is a symmetry transformation. The significance thus plays a discriminating role, similar to that of energy operator H of the quantum structures: the comparison between meaning and Ω must prove unproductive to be able to classify Ω as a conceptual symmetry transformation.

In other words, keeping Lee's above-mentioned statement in mind (p. 58), *one is inclined to recognize symmetry every time one is not aware of having made a structural transformation*.

Metrics: where the rhythm rises almost to the role of significance

Thus, as we saw in metrics, a translation of the feet is symmetric (in time) as long as it doesn't produce a perceptible alteration in the rhythm. Analogously, the scansion of the concepts evoked by an image is symmetrical if it doesn't cause a perceptible change in them. For example, Fig. 29 can be considered symmetrical as long as the observer continues to see only a young woman or only an older woman (but let's not forget that "life begins at fifty!"). Finally, a face may appear symmetric up to the moment in which we note a different expression in the right eye than in the left eye, etc.

What counts in the conceptual synthesis therefore isn't so much the energy as the information, which is the quality of the energy. At the level of conceptual synthesis, a structure appears qualitatively symmetrical until the moment in which, following a scansion of the structure itself, a dynamic instability of the perceptive process is produced: and one realizes that he has extracted information, or given a new meaning, or has been enlightened by an idea.

Breaking the sym · metry then is equivalent to shattering a com · mensurate equilibrium of structural relationships of a dynamic, geometric, rhythmic, acoustic, conceptual, etc. nature.

When symmetry is insured in an ambiguous figure to give the pleasure of breaking symmetry

In the artistic field, inserting elements of symmetry into a structure is a primordial, aesthetic need. Elements of symmetry are incorporated in a work of art or in poetry more or less latently: on the other hand, this is the way the artist is able to give himself, or perhaps with the observer's active participation – as occurs when perceiving ambiguous structures – give, the pleasure of breaking symmetry.

At this point we should offer a few examples.

An ear trained to the six-six, or four-four-four, or three-three-three-three rhythm of the Alexandrine, the twelve-foot French verse (Caillois, 1973), doesn't miss the breaking of symmetry in the five-seven line: a scanning with which Victor Hugo enlivens the tale of a general on horseback who, after the battle, meets a wounded enemy soldier and offers him a flask of rum. The soldier, instead of thanking him, shoots at him:

6		6
le coup passa si près	|	*que le chapeau tomba*
et que le cheval	|	*fit un écart en arrière*
5		7

In the field of music, one could propose the "conflict between background and form" in the paroxysmal sequence (Plate X) of the Prelude in Partita III

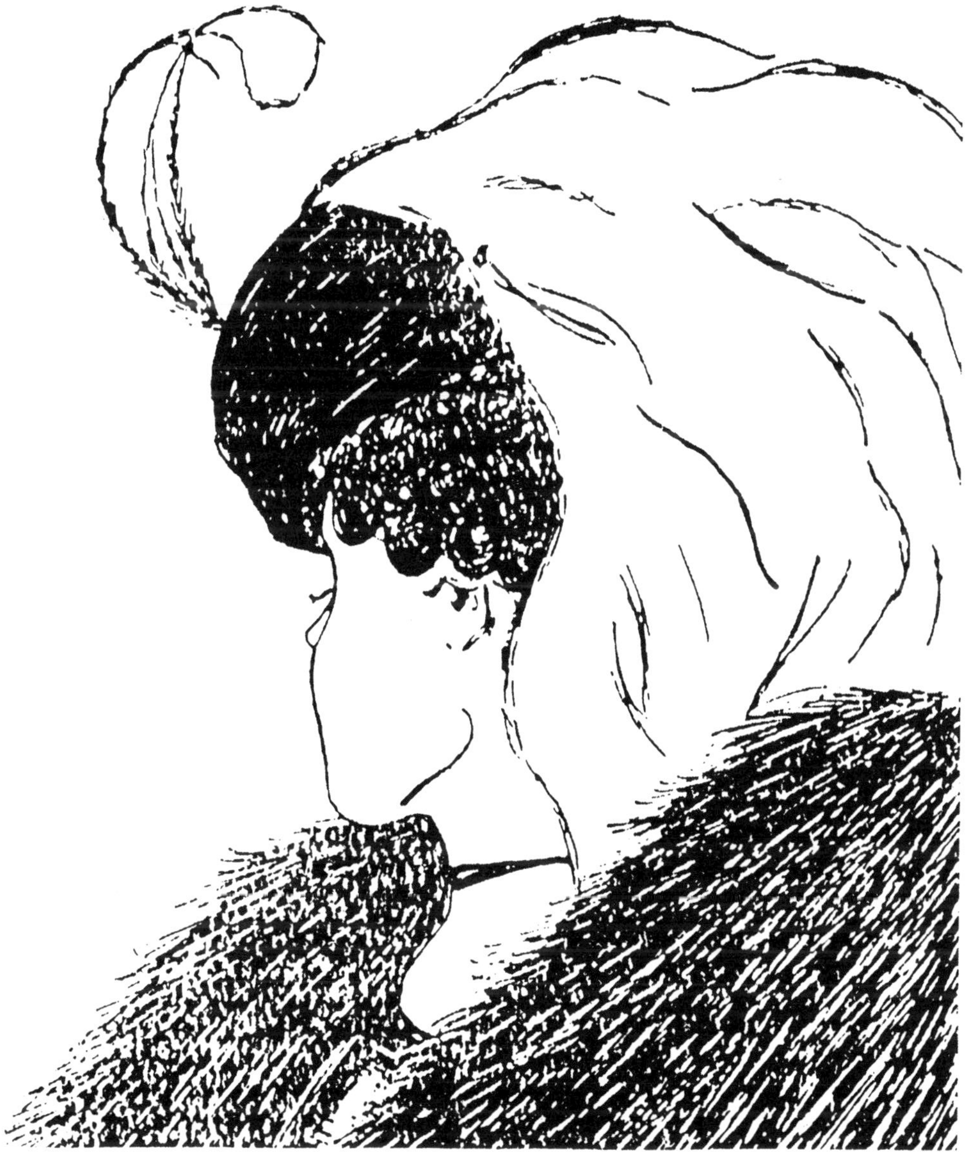

Fig. 29. Illustration representing a young woman and an elderly woman, by the charicaturist W. E. Hill (1915). The young woman's chin is at the same time the older woman's nose. (From F. Attneave: Le Scienze, no. 43, March 1972, p. 84)

for violin solo, which J. S. Bach, "a genius of the baroque as well", invites us to compose.

The dynamics of ambiguity and the breakings of symmetry

Our auditory thinking is stimulated to break the translational symmetries in which, at a synchronous reading, the resonant modular structure of the score seems to abound: so that one feels rewarded when, after intense and repeated effort, he manages to listen to the dynamics of the ambiguity that springs from these breakings of symmetry.

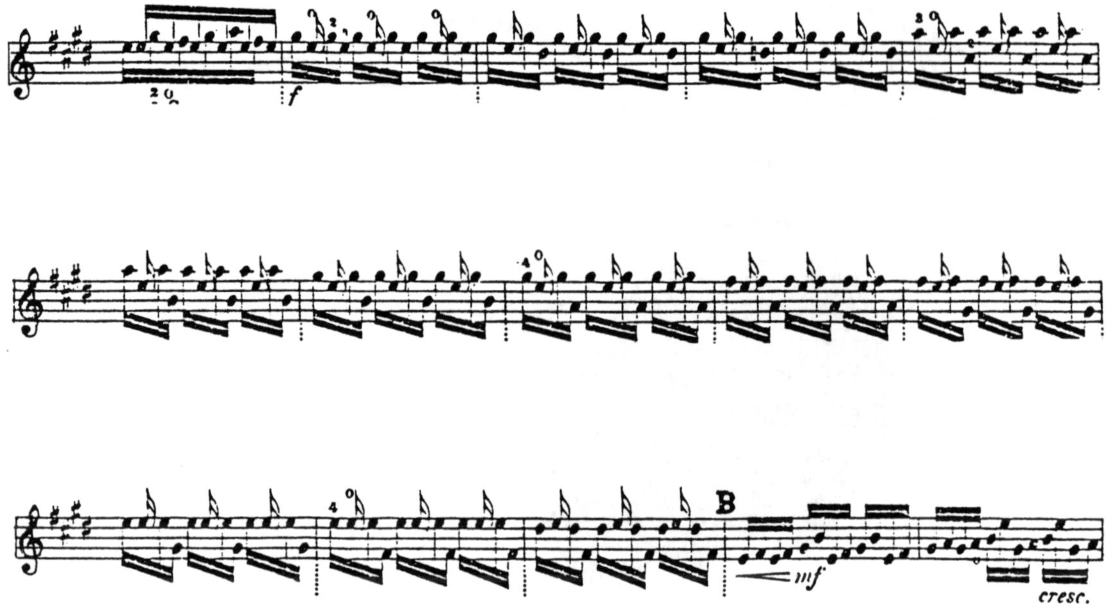

Plate X. Fragment of the Prelude from Partita III for violin solo in E major by J. S. Bach, BWV 1006.

Starting with the "forte" on the second beat – which is really the seventeenth in the prelude – and up to B, the E of the empty chord appears systematically every four demisemiquavers, in positions 2, 6 and 10 of each beat. The musical structure, precisely because of this repetitivity, is thus characterized by translational symmetry, albeit partial, at a rate equal to a fourth. From the point of view of rhythm, the above-mentioned positions – even – are secondary, since the accent would naturally fall on the positions – odd – 1, 5 and 9. Nevertheless, the sound of the empty chord is rich in higher harmonics and more intense, so that it tends to seize the attention and capture the accent. The listener then faces a choice between two possible sequences of accents (even or odd): he is surprised to experience, at the auditory level, a sensation similar to the one he feels when dynamically perceiving, at the visual level, the ambiguity of Schröder's stairs (Fig. 4) or the wall of the église d'Auvers (Plate II). In simpler terms, try pronouncing aloud the sequence

. . . *delaydelaydelaydelaydelay*. . .:

a slight hitch in placing the accents causes one to pass sooner or later from the "principal melody" (*delay*) to the "secondary melody" (*lady*).

In the visual field, breakings of symmetry introduced directly into the painting are frequently found in Giotto. A traumatic innovation in the context of figurative Byzantine art, they are incorporated into the structure and, unlike what occurs in the perception of the ambiguous structures, they are not brought on by the observer's active control of the structure itself. They produce the effect – inverse, in a sense – of offering themselves as regions on which the observer's control focuses instinctively: think for instance of the breaking of translational symmetry in the sequence – consignment of the staffs, Mary's wedding – in the Scrovegni Chapel at Padova; or of the face and hands of Saint Peter at the center of the predominantly symmetrical structure of the

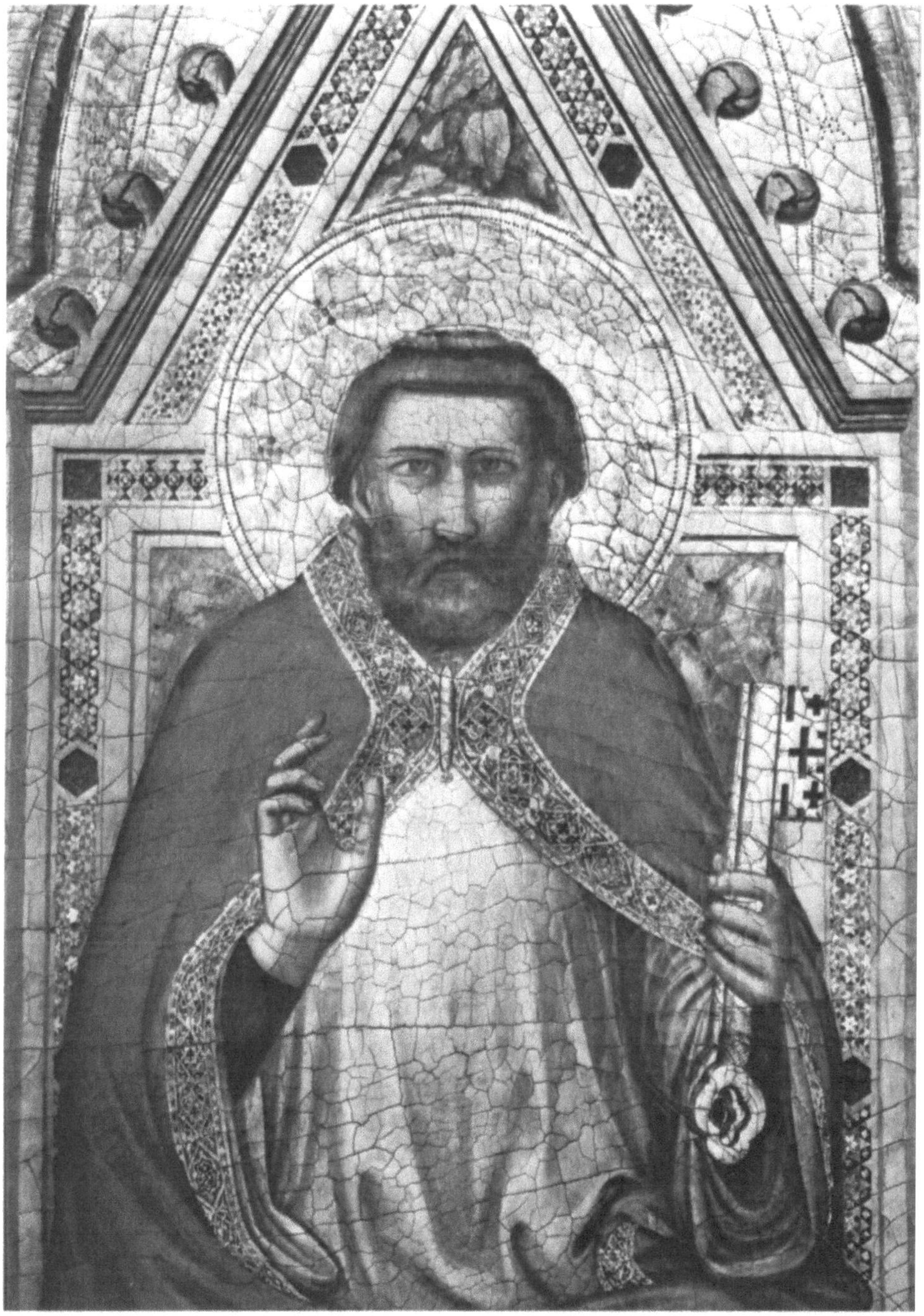

Plate XI. Bottega di Giotto. Saint Peter Enthroned. Polittico Stefaneschi, Vatican Gallery, Hall 1

throne by Polittico Stefaneschi (Plate XI): the Saint's restless and slightly cross-eyed glance attracts the observer irresistibly and at once becomes the pole of his visual thinking.

Finally, in poetry or the visual arts a structural transformation, in order to be classified as a *symmetry transformation*, must not disturb the essence of the structure. In particular, it should not noticeably modify the structure's significance. Or, at least, it *must preserve* or not compromise *the existence of the significance* itself.

Symmetry as commensuration of relationships among structural moduli

Thus sym · metry implies com · mensuration of relationships among structural moduli, which in its turn is the result of the imperceptibility of differences among the moduli. In this com · mensuration of relationships among moduli intrinsically in equilibrium – an equilibrium which so often is begging to be broken – we thus rediscover, along with Weyl (1952) the fascination of an "idea that has guided man throughout the centuries to the understanding of order, beauty and perfection": (sym · metry) *συμ · μετρία.*

Could my friend the Hellenist Raimondo Malgaroli be right when he doesn't miss a chance to remind me that "the Greeks have already said it"?

Insert

The Musicality in Virgil

> The scientific process is the process of scientific objectivization where objectivization is understood as a closer approach to Truth. It is thus essentially cognitive, although retaining extra-cognitive aspects in its essence (...). The artistic process is the process of artistic objectivization, where the objectivization is understood as a closer approach to Beauty. It is therefore essentially aesthetic, although containing extra-aesthetic aspects in its essence.
>
> Joseph Agassi, 1979

Virgil's poetry

The curtain has fallen on the celebration of the second Virgilian millennium,

mā́ĭŏ|rḗsquĕ că|dū́nt| |āl|tī́s dē | mṓntĭbŭs | ū́mbrāē

While the tall mountains cast lengthening shadows of the night that separates us from the dawn of the next millenium, seeing as how there's time, permit a non-expert in the field to enlarge upon an element of Virgil's poetry that apparently lends itself to being dealt with at a not-exclusively literary level: its musicality.

Poetry and sound

Montale writes: "if I take poetry as an object, I believe that it is born of the need to add a vocal sound (word) to the beating of early tribal music (...) In the first tales of the Nibelungs and later in the romance tales, the true nature of poetry is still its sound".

A problem arises. What do Virgilian hexameters have in common with the beating of early tribal music? Is sound really so important, even in Virgil's poetry?

Metrics and song

In Virgil, sound, metrics and song are adapted each time to expressive needs, so that form and content often turn out to be the same.

But one thing at a time. The snares that our domestic reformers have laid for Latin in school make – I'm afraid – a digression on meter necessary. Don't worry: a few words will suffice; we already feel certain ideas in our bones.

The hexameters – like the one transcribed at the beginning of the chapter – are composed of six "meters" or "feet". One foot contains two or three syllables, and thus the same number of vowels (note however that even for the ancient Romans, a *u* following *q* was not considered a vowel. Moreover, the citizens labelled S.P.Q.R., who descended from the Ara *coe*li as the umbr*ae* of the evening fell, assimilated the dipthongs – the pairs of consecutive vowels *ae*, *oe* – into a single "long" vowel). There are various types of meters: dactyls, spondees and trochees; they are represented by the symbols ⊥⏑⏑, ⊥—, ⊥⏑, respectively. The sign —, placed above a vowel, indicates that it is long: in pronouncing it the voice must draw it out. The sign ⏑, placed above a vowel, indicates instead that it is short: the voice doesn't linger over the pronounciation. Long vowels (or dipthongs) are the only ones that may be accented. For example: the first *a* in Italia in Mameli's anthym. Or the dipthong *eu* in

Beethoven and Archilocus

Freude, right at the beginning of the hymn to joy in Beethoven's ninth, which – we notice it at once – explodes with the rhythm of the catalectic trochaic tetrameters, to celebrate the adhesion of Greece (Archilocus' fatherland) to United Europe.

As we were saying, hexameters are composed of six meters. The first four are dactyls (⊥⏑⏑, the sign ⊥ indicating a long accented vowel) or spondees (⊥—), the fifth is usually dactyl and the last is trochee (⊥⏑) or spondee.

Assuming that a dactyl always be found in the fifth position (Virgil isn't Catullus!), let us concentrate on the first four feet. Each of these may be a dactyl (d) or a spondee (s). If the sequence of possible binary choices between d and s were purely accidental (think of heads or tails on a coin) we should find one composed in all of 5 d and the final spondee or trochee, on the average of every sixteen hexameters ($2 \times 2 \times 2 \times 2 = 16$).

Virgil and Tartaglia

A student of musical semiotics who is interested in playing the lottery and perhaps in late numbers, could find the distribution of d's and s's in the first four feet of all Virgil's hexameters and compare it with that expected for the binomial coefficients, according to Tartaglia's triangle: in other words, if the poet's choices were entrusted to chance, in a group of sixteen hexameters, on the average, the first four feet should present the following recurrences (see table in Fig. 39, especially the line marked by $N = 4$):

once, as we said, four d's;
four times a d and three s's;
six times two d's and two s's;
four times three d's and an s;
once four spondees $(1 + 4 + 6 + 4 + 1) = 16$.

In the bucolic poems, none of the eighty-three verses in the first eclogue is pentadactylic. But this lack, which has perplexed me since I was in high school, is compensated for in the subsequent eclogues.

In the eighth eclogue for instance, the same pentadactylic hexameter is repeated like a refrain, with the anxiety and insistence of a lover, a good eight times (after all, "God delights in uneven numbers"):

"**Du**cite ab |**ur**be do|**mum**, mea | **car**mina, | **du**cite | **Daph**nin",

bring him home again from the city, my songs, bring Daphne home.

Whoever "with the tapping of a finger" follows or recalls the numbers dispersed in Virgil's verses can find an almost systematic correlation between the pentadactylic hexameters and the dynamic idea of movement they evoke.

In this regard, that

quadrupe|**dan**te pu|**trem** soni|**tu** quatit | **un**gula | **cam**pum

The onomatopoeia

from the Aeneid is symbolic (it should be translated retaining, more than the sense, the onomatopoeia: dust and turf are raised by the hoof and the pounding resounds). Here the evocation of the quadruple sound with which the galloping horse's hoof hits the dusty field is the fruit of choice and certainly not accidental.

What is the segret that permits Virgil to call up the rhythm that animates this resonant structure?

Montale gives us permission, so let's look at this hexameter like an object. And why not, like a Corinthian cup (Plate XII) on which the horse just happens to be represented. Let us observe one of the horses in the upper part in particular. One notices immediately that it is possible to make a transformation on the vase without being aware of having done it: this is a rotation by steps, which has the effect of translating the horse behind, having it advance until its position coincides with that of the horse in front. In this way a transformation is realized that operates on the vase in such as way as to keep it unchanged: Such a transformation, which operates on a structure without altering it, is a symmetry transformation: the representation of the horse on the Corinthian cup possesses translational symmetry. Like every symmetry operation, translational symmetry constitutes a reference for the structure and concurs in determining its significance.

Translational symmetry and the rhythm

In physics the operation which generates translation is the momentum. If translation is a symmetry operation however, one does not notice the variations that may occur as a result of the translation. In particular, the translation does not bring about changes in the energy of the structure. The translation takes place therefore as if the structure were isolated, not influenced by external forces. Under these conditions, the structure's momentum is constant. After all, this is the reason why whoever painted the Corinthian cup, not by chance adopted this modular scansion of space in depicting the horse, to give the idea of regular and rhythmic movement.

Let's return to the pentadactylic hexameter, transcribed in the form of a pentagon in Fig. 30. Leaving aside the final campum, we could treat the dactyls by the same standard as the equestrian moduli that form the vase. Let's see if we notice an alteration of the verse resulting from a translation of the dactyls that closes on itself, like the translation of the equestrian moduli made on the vase. That is, let's substitute the first dactyl with the second, the second with

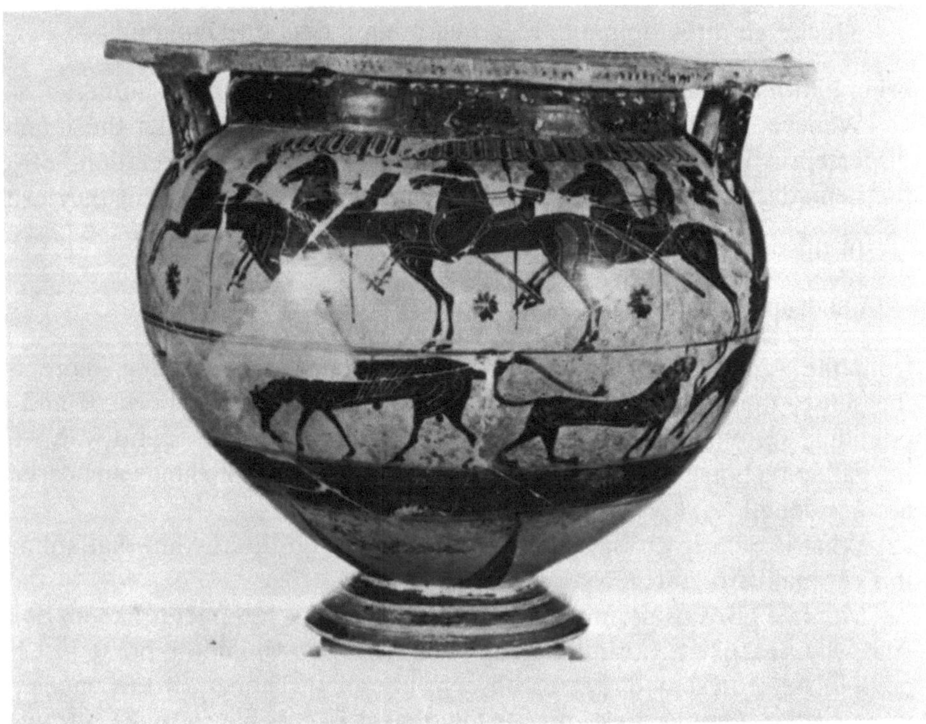

Plate XII. Corinthian cup: Horsemen with animal frieze. Rome, Archeological Service for Southern Etruria, Villa Giulia Museum

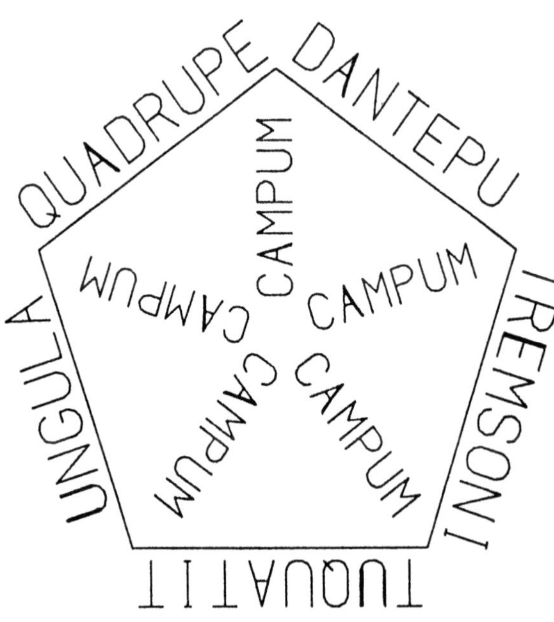

Fig. 30

the third, ... the fourth with the fifth and, to close, the last with the first. We obtain – ah, Dante, what an insult to your Virgil –:

D**a**nte pu|tr**e**m soni|**tu** quatit | **u**ngula | qu**a**drupe | c**a**mpum.

Dactyls and horses

As a result of the translation the rhythm, one perceives, remains unchanged. Because of the translational symmetry in the structure of the pentadactylic hexameter, the rhythm becomes a reference for this poetic structure and contributes to determining its significance. The translational symmetry which in a poetic structure leads us by the meter toward an essentially periodic scansion is associated, as we mentioned, to the concept of constant momentum.

After all, this is the reason for which Virgil, not accidentally, gives translational symmetry to his hexameter when the predominant significance of his verse lies in the periodicity or repetitivity of an action, or in the rhythm: in this way he evokes, onomatopoeically, the constant momentum, that is, the rhythm.

Up to this point we have discussed the rhythm, or if one prefers, the beating of tribal music. But the accompaniment of a battery is not music, and rhythm, by itself, is not poetry. For music and poetry, sound is needed.

Let's try re-reading aloud the final verse of the first eclogue, transcribed at the beginning, scanning it the way our old Latin professor would have liked. It won't be long before we become aware of a contrast among the sounds, which is in harmony with a dynamic rippling of the rhythm.

Acceleration and symmetry breaking

Regarding the sounds, that acute central *i* of the verse – a final flicker of light breaking in after the caesura – flashes between sombre *u* and the *o* of the third dactyl: these are all solemn vowels, that lengthen deep as the shadows of the evening. Regarding the metrics, the dactyls alternate with the spondees: consecutive accelerations of the rhythm result, and the translational symmetry is broken.

Finally, we may conclude that musicality is one of the channels through which Virgil, harbinger of the baroque, transfers the essence of nature – whose laws he instinctively divines – into his work: laws which he then expresses heightening greatly the evocative power of his song.

Insert

Exchange Symmetry

> "Perhaps the most essential difference between objects in the macroscopic world – common objects, that is – and objects in the microscopic world of the atoms is the following:
>
> In the macroscopic world one never finds two objects that are equal. Consider two pieces of iron, for example (...), formed of groups of billions of billions of atoms: if just one of these atoms in one of the two pieces of iron is displaced with respect to the corresponding atom in the other piece of iron, the two objects can no longer be called identical. In this sense therefore, the non-existence of identical bodies in the macroscopic world may be interpreted as evidence of a highly complex structure.
>
> The situation becomes quite different from this point of view if, from considering common objects, we go on to atoms and molecules, or better still, their components, nuclei and electrons. In the world of atoms in fact one frequently finds objects that are identical to each other: thus, for instance, it may be that any two electrons, or even any two atoms of the same kind, are identical (...). The great precision with which the identity of two electrons may be established depends on the possibility of observing the effects of a possible difference accumulated over a very long time. For example, if there were even a slight difference between two electrons contained in the same atom, the effect of this difference would also be slight, if observed over a brief period in the atom's life. But in the long run it would end by changing the structure and the external properties of the atom in a fundamental way. Therefore we may affirm the identity of the two electrons, if not in an absolute sense, at least within exceptionally reduced limits of precision."
>
> Enrico Fermi, 1933

Symmetry as a reference

The concept of symmetry is associated to the property, typical of symmetric structures, of remaining unchanged as a result of transformations. It is therefore a reference point in the study of nature.

In the twenties, when the structure of the atom was still a subject on the frontiers of physics research, it was realized that in atomic structures and natural structures in general, together with symmetry transformations of a geometric nature, another class of symmetry transformations must be examined: specifically, those which consist in the exchange of identical or indistinguishable particles.

Identical particles

The exchange symmetry

Think of the atoms. The electrons "orbit" around the atomic nucleus, but there's no way to distinguish one electron from another. Even by making them collide like two identical billiard balls, one could never be sure that at the moment of collision the incident electron and the target electron were not exchanged. Exchanging electron A with electron B and electron B with electron A leaves the two-electron system unchanged: the electrons escape our every attempt to baptize them. Consequently, the exchange A→B and B→A is an indiscernible transformation; it's a symmetry transformation.

The exclusion principle

On the other hand, nature requires that it be impossible for more than two electrons to be hosted in a single physical state, or as we might say, in the same cell of the six-dimensional space of positions and momenta. In other words, if two electrons are in the same position and have the same momentum, it is *excluded* that their position and their momentum coincide (even within the limits (of uncertainty) fixed insuperably by the uncertainty principle) with the position and the momentum of a third electron. This is Pauli's *exclusion principle*.

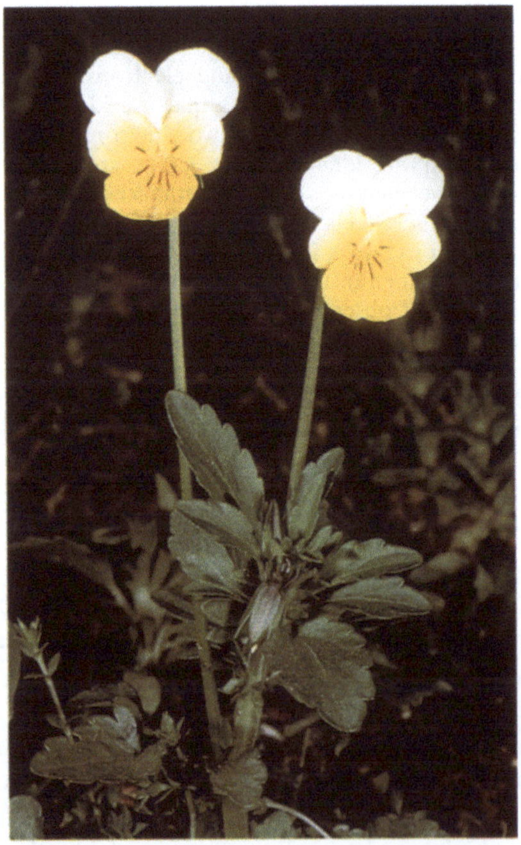

Plate XIII. Two twin sweet violets two (photo by Mario and Umberto Parravicini)

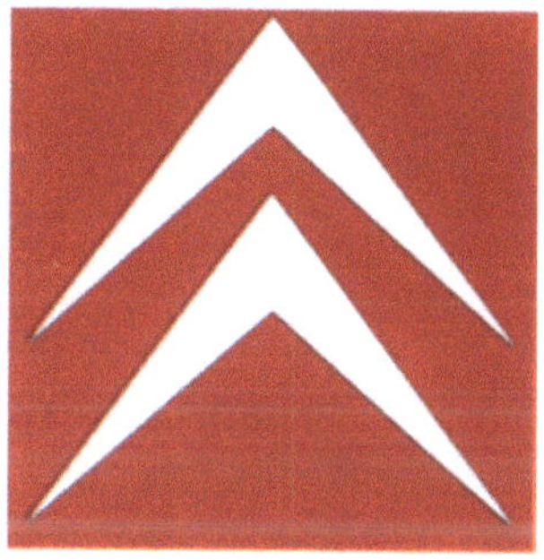

Fig. 31. The Citroën trademark. The exchange symmetry, at the basis of the material's structural characteristics and stability, is revealed in the indistinguishability of the twin moduli. It instils a feeling of reassurance whose roots reach deep into the unconscious. This feeling relieves the sense of unease that the sharp points on the ends of the moduli forming the trademark produce

Fermions

This principle applies only to the states of certain identical particles called *fermions*: electrons and other equally unsociable particles, which were called elementary before quarks became fashionable: neutrons, protons, etc.

Bosons

It should be said however, at least in passing, that other identical particles, called *bosons*, have the opposite tendency: instead of excluding themselves, the bosons, sociable beyond belief, tend to crowd themselves all into the same state. An example: the monochromatic photons of laser light, collimated and in-phase – the spear of colored light, darting boldly around the discotheque.

Let's get back to the electrons. If the electrons were distinguishable they would ignore each other. The exclusion principle would not apply to them. The exclusion principle thus rests on exchange symmetry.

But the entire world around us rests on the exclusion principle.

Atoms are made as they are because electrons organize themselves around the nuclei in compliance with the exclusion principle; and precisely because this principle is respected, the electrons give rise to the periodicity characteristic of the structure and properties of atoms that Mendeleev pointed out, on empirical bases, a century ago. The sequence "lithium-sodium-potassium-rubidium-caesium", like an alkaline nursery rhyme, comes to mind. Nature is made as it is because the electrons, which in the chemical bond act like glue among the atoms, organize themselves in pairs, following the exclusion principle. Magnetism, superconductivity and the Hall quantum effect – a discovery which won K. von Klitzing the 1985 Nobel Prize for Physics – are macroscopic manifestations of the exclusion principle.

Aesthetics and the exchange symmetry

Even our human structure rests on the exclusion principle. To the point where it might be the case to ask ourselves whether the aesthetic sensations and emotions we experience in the presence of identical – or nearly – structural moduli in works of art throughout the ages, in nature (Plate XIII) and in design, aren't simply the reflection of structures, dynamics and logics that the exclusion principle and, preceeding that, exchange symmetry have imprinted once and for all in our brain.

Insert

The Game of Broken Symmetries in the Industrial Trademark

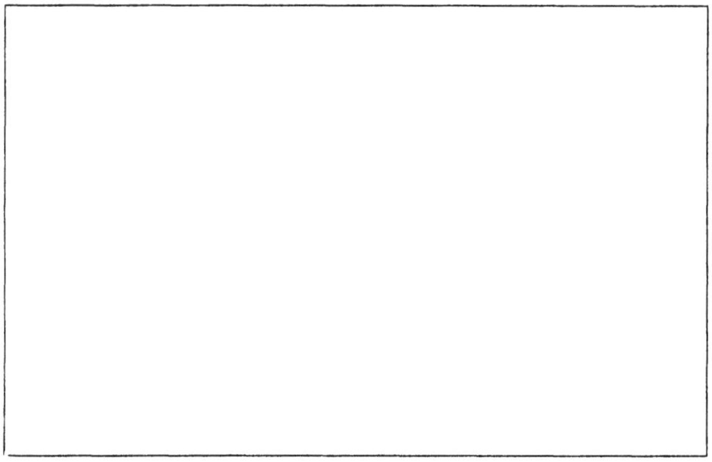

Fig. 32. The white cloth and symmetry: in its indiscernibility lies the significance

The consumption of energy, promoting the consumer society and the information society

In all industrialized countries only a modest portion of the energy consumption is used to maintain the population. Actually, almost all of the available energy is given over to manufacturing and utilizing industrial products and to the organization and management of services. And the energy introduced per unit time into the socio-economic system assumes the role of "control parameter" of the system itself.

Thanks to energy, products and services as a whole tend to expand rapidly. Thus energy becomes a source of options, while our existence is enrichened by stimuli and meanings. In this increasingly complex and intricate world, external pressures to make choices and the internal need to remove uncertainty or create information are urgently felt.

This is how the consumer society and the information society are born.

The multiplication of new products and services and the need to handle an excess that appears necessary to us by now, lead to enrichment of the language and development of non-verbal forms of communication.

Language as social order parameter

Until several years ago, language was a more or less unchangeable point of reference: it underwent almost imperceptible variations during an individual's lifetime and became – to use Hermann Haken's words – an order parameter which enslaves those who use it. But lately the rapidity with which the lan-

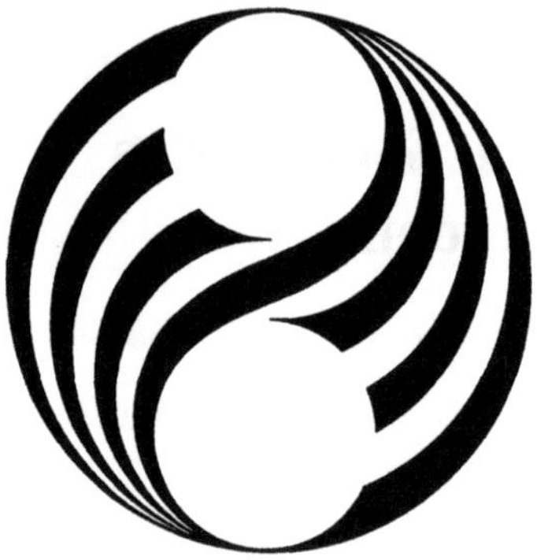

Fig. 33. Franco Grignani: the Molinetto Tennis and Golf Club trademark. This trademark recalls dynamically the gestures of both golf and tennis players

guage has been adding on new words and ridding itself of archaic ones seems to have increased suddenly. This swift revolution of the language causes confusion: an order parameter, like any rule, does not generally vary a great deal during the time required to carry out the elementary actions that come under its jurisdiction. Consequently it's natural that when an order parameter, like language, varies considerably over a period that is relatively short with respect to the cultural inertia of the social system, it should produce disorientation.

Images and the language of visual communication

The psychological void created by the flood of new words into ordinary language tends to be filled by the production of new images, which enrichen the language of visual communication.

It is then that an image culture explodes. This happens not just because images can be easily prepared, diffused and reproduced using the technology of informatics and telecommunications, but also and above all because, without having to be translated, they offer themselves directly to the imagination of whoever perceives them, and they strike his sensitivity. Easily impressed on the mind, the perceived images tend to correlate themselves with acquired experience, and to re-evoke it; plugging themselves directly into the observer's psychological sphere, they provoke an abrupt reorientation of thought, resulting in an involvement that is emotional as well (Fig. 33).

New words and images, then.

New words and new images

New words in the fields of research, development and production. New images in marketing and distribution.

In these two important sectors of economic activity, images prove to be more effective than words when it comes to extolling new products, services and performance characteristics to potential clients. At times the words themselves become images. And ordinary language and visual language, supplementing each other, end up by sharing the role of order parameter.

A hidden persuader: the trademark

Among the images in the technological world, the industrial trademark occupies a privileged position. A fundamental mission has in fact been entrusted to this omnipresent "hidden persuader": to induce the consumer to choose one product or one service over other products or other services offered by the competition, helping to direct his choice.

We may ask ourselves if it is possible to define the criteria for designing a trademark that will perform this specific function.

Actually, the synthesis of a trademark is the result of a risk taken at the meeting point between psychology of choice and perceptive mechanisms, between the art of persuasion and the science of communication.

Thus the question that we have raised is not an easy one to answer. But we will try.

Let us note preliminarily that to be effective, a trademark must be constructed so that it acts as a stimulus that may be syntonized on the observer's – the potential consumer's – psychological sphere.

Having said this, we should analyze the interfacing between the structure of the visual message suggested by the trademark and the observer's typical state of mind, as it develops during the critical phase of choice: the more consistent the interface, the more effective the trademark will be.

How to structure a trademark in order to make it more effective

On one hand, then, we have the trademark, with the structural and potentially dynamic characteristics belonging to the system of signs that compose it; on the other we have the observer, with his impressionable state of mind which moves from a basic, initial uncertainty, and through a dynamic instability, evolves toward the final decision.

For the trademark to be effective, therefore, it's not enough that it – though using a universal message that has no need of translation to be received and impressed in mind – illustrate the salient characteristics of a product, a service or a company. The composition of the trademark, interiorized in the observer's mind, must also be or become so much a natural part of his mood that it emerges spontaneous and dynamic at the critical moment of choosing.

Symmetry and reassurance

The consumer, with a pressing need to rid himself of uncertainty, explores at length, and from all angles, the many options open to him: he weighs them, looking for that "quid" that will enable him to make a choice. Instinctively he lends his attention to whoever takes the trouble to inform and persuade him. With a tendency to be suspicious, he is sensitive both to fixed and repetitive references calculated to generate trust (Fig. 34), and to dynamic elements calculated to reveal favorable perspectives (Fig. 35).

The consumer after all craves reassurance and incitement at the same time. And he responds gratified to any stimulus (of a graphic nature perhaps) in which the expressions that best represent his mood's contradictory needs converge.

He finds reassuring references in those trademarks containing elements of symmetry: center or planes of symmetry, rotation axes or lattices, so that overturning or reflecting, rotating or translating it, the structure of the trademark remains exactly the same, and one is not aware of having done so. But at the same time, he grasps with relief and curiosity the presence of elements

Fig. 34. In the Mercedes trademark the rotational symmetry – associated to the fact that each third of a complete turn keeps the trademark exactly as it was – evokes a uniform movement, which occurs with constant angular velocity. This is a reassuring reference for the observer: no surprises coming up!

Fig. 35. The Renault trademark also appears to be symmetrical at first. But at the moment when the observer actively breaks the symmetry of this figure, perceiving its ambiguity, a dynamic element arises vivaciously which seems to allude to the opening of new perspectives

Fig. 36. The FIAT trademark offers a reassuring reference through the uniform scansion of the plane, obtained by repeating the rhombic moduli with translational symmetry. At the same time, with the breaking of symmetry brought on by the letters of the time-honored Torinese acronym, the trademark seems to herald musically the opening up of new perspectives

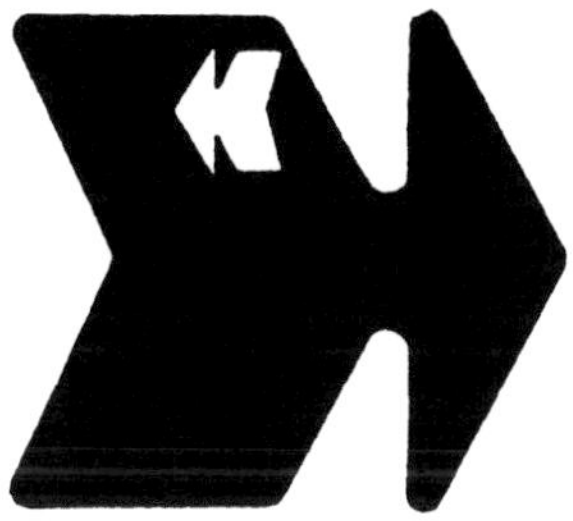

Fig. 37. Luciano Consigli: trademark for the journal "humor graphic". A point of irony – the point, in fact, of the white arrow – is all that's needed to break the symmetry and the hypnosis, the drabness and monotony of dull and uniform routine: countercurrent, biting and incisive

Broken symmetries and incitement

which can break the repetitivity, the commensurate monotony and the uniform homogeneity of the equilibrium typical of every symmetric structure, and trigger, by upsetting this equilibrium, dynamics that may evoke crises, changes and new perspectives (Fig. 36).

To conclude, two points:

First. It doesn't seem rash to recognize in the careful and calibrated dosing of symmetry and broken symmetries a fundamental criterion for designing a trademark that will be effective and unmistakable among the idols of the technological Olympus (Fig. 37).

Second. The analysis of the trademarks offers ample material for reflecting on the bases of graphic language: a language that we should learn to understand bctter so as to better understand the increasingly complex society in which it develops.

Insert

Selfsimilarity in Nature and Culture: The Fractals*

Among all the forms of symmetry there is one that is typical of many microscopic and macroscopic structures, both inorganic and biological: selfsimilarity. *Selfsimilarity*

A selfsimilar structure is characterized by its likeness to itself in each of its parts (Plate XIV).

A typical example of selfsimilar structure is the fractal. The word 'fractal', a recent addition to the dictionary, was coined in 1975 by Benoit B. Mandelbrot, a mathematician of Polish origin, professionally formed in France. He was able to give a concrete and familiar physical significance to mathematical ensembles born of the creativity and imagination of scholars such as H. Poincaré, G. Cantor, G. Peano, F. Hausdorff, G. Julia and P. Fatou. To these ensembles, usually defined by jagged, broken and irregular lines or surfaces, Mandelbrot gave the name 'fractals'. *Fractal*

Beginning with these aggregates, which at the time were considered "monstrous" or "exceptional" and would still be buried in algebraic and esoteric form in dusty archives today, Mandelbrot developed, classified and presented in a form even visually accessible and usable, a corpus which now offers itself as an alternative to nothing less than Euclidian geometry. According to Mandelbrot, in fact, circles, triangles and the other geometric figures of Euclidian geometry, although very dear to Galileo, are not of much help in describing natural forms. Classical geometry is "cold" and "dry" because "it's incapable of describing the form of a cloud or a mountain, a coastline or a tree. Clouds are not spheres, mountains are not cones, coastlines are not circles, treebark is not smooth, nor does lightening travel in a straight line". *Euclid and Mandelbrot*

What, then, is a fractal? A fractal is a structure each part of which is a copy of a part of the structure, and in this case – rather like the Russian dolls – it is a copy of the whole structure: a fractal therefore is a selfsimilar structure.

Each segment of the agave flower is a fairly faithful copy of a segment of the agave flower. In boiling water, the steam phase and the liquid phase compete for the water: thus, at equilibrium, the dispute is settled and a critical configuration is realized in which a lake of steam contains islands of water drops, each containing a tiny lake of steam which contains islands of water

---

* This insert is based on a paper by the author published in *Modo*, N 113 (April 1989), pp. 58–59.

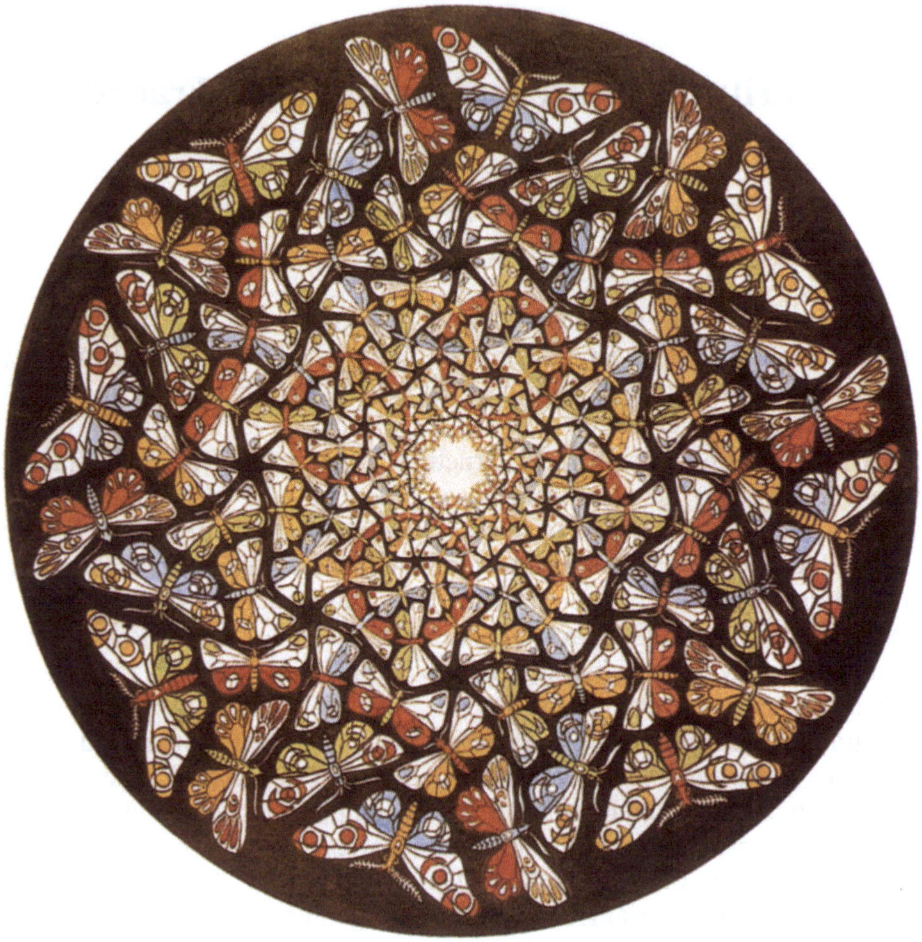

Plate XIV. Butterflies. Watercolour by M. C. Escher. (©1950 M. C. Escher/Cordon Art, Baarn, Holland)

drops inside each of which ..., in a zooming that brings to mind the infinity in which the gaze of two lovers who discover their own eyes in the eyes of the other, is lost.

The beauty of fractals lies precisely in the indifference or invariance of their form with regard to the enlargement process. Man is perhaps the most curious animal. He needs to learn and know, and to know and learn he must implement or strengthen the sharpness of his senses. By nature, then, man likes to enlarge. Galileo pointed his telescope on the stars, while today's physicists, using the tunnel-effect microscope, are able to distinguish one atom from another, and with increasingly gigantic particle accelerators explore the smallest details of the structure of the atomic nucleus and the "glue" holding it together. When, by dint of enlarging a detail of a fractal we return, after an orgy of images, to an image identical to the initial one, we feel relieved: it's a little like returning home after an interesting but tiring trip. The fact that the form of an image does not alter when it's enlarged is reassuring – just as the unchanging rhythm of a litany or the monotonous refrain of a lullabye is reassuring.

What then is the mechanism or rule on which the invariance of fractals when enlarged is based? It's an iterative rule, a recipe, which establishes the way in which, starting from one point on the plane, another point on the plane can be reached; or starting from the point reached in this way, one arrives at still another point, and so on. It is precisely the regularity with which the rule is applied that is transferred and manifested in the scale invariance or self-similarity typical of fractals. For the rest, it's a simple rule: to apply it on the computer, only few lines of program are needed.

The rule governs the mapping out of the plane, or rather the dividing of the plane into basins of influence centered on singular points called attractors. This controls a kind of topological dispute, which consists of a dense sequence of choices that dynamically depict a critical process. Finally, it's an incredibly fertile rule, which generates a galaxy of images.

As predicted by H.O. Peitgen and P.H. Richter in their "The Beauty of Fractals" Images of Complex Dynamical Systems (Springer, Berlin 1986), the development, in recent years, of this sector of physics, mathematics and numerical calculus has widened the range of expressive means, enriching linguistic structures and facilitating the understanding of many natural and cultural phenomena: the development of fractals, then, marked a turning point in the discovery and development of innovative expressive forms in design, art and visual communication.

It is certain that whoever explores the world of fractals is immediately struck by the fascination of these images – images which, linked as they are to the dynamics of the decisional process, offer our minds not so much archetypal forms as simple archetypal rules on which the formation of the forms is based.

Simple rules and complex forms

A careful examination then leads to further surprises. It seems incredible that extremely simple mathematical rules (although non-linear) make it possible to generate an unlimited variety of such rich and complex forms. And it seems incredible that viceversa the complexity and multiplicity of so many forms may spring from algorithms apparently so lacking in content. Man has always hoped to trace the complexity of reality back to first principles. The geometry of fractals appears to be a natural candidate for supplying representations and simulations of natural structures. And the growth of knowledge in this field leads us to predict that this geometry may prove to be decisive in furthering man's age-old aspiration.

There is something biological about fractals. Just as genetic information is contained in each cell of the body, information on the whole structure is compressed in every part of a fractal. After all, when the numerical values associated to a fractal are set so that it may be constructed on the computer, it is usually impossible to know beforehand how the creature the program produces will look. It will however seem perfectly beautiful to the person creating it, like a baby to his parents.

Creativity and ingeniousness

In concluding, it should be suggested that in terms of fractals or the way they are generated, we might perceive the dynamics through which not only the manner of producing natural structures, but even creativity and ingeniousness in art is expressed. The creativity and style of a Mozart or a Bach could stem from their ability to identify and apply with great speed the rules of a game that is theirs alone: the rules – characteristics of their style – of a game structurally analogous to that which makes it possible to create mechanically, as in a kaleidoscopic bolero, countless fractal images starting from a simple, repetitive prescription. But the brilliance of these musical 'greats' lies almost certainly in the sixth sense with which they know how to grasp the right moment for breaking, with expertise, the rule of this iterative game and adopting a new one.

Chapter Six

Conclusions

> In the urge to look for the way out, there is always a certain amount of love for the labyrinth in itself;
>
> And a certain zeal in looking for the way out plays a part in the game of losing oneself in labyrinths.
>
> Italo Calvino, 1962

The final chapter in a well-constructed study should open by listing the subjects discussed, present a comparison between the final balance and the objectives set at the beginning, and close by indicating perspectives that further development of the theme would offer.

To respect this criterion while reducing the necessary space to a minimum, I would suggest that my twenty-seven readers review the index and the inserts, glance over the introduction and the first chapter, take another look at the opening part of the following chapters and the bibliography, the figures and the plates and then attempt, autonomously but – for heaven's sake – without overdoing it, to evaluate how well the finished product meets the objectives.

For my part, although the temptation is strong, I will refrain from suggesting implications and extrapolations of the topics discussed. For two reasons. In the first place, any possible extrapolation, for instance into philosophical, psychological, ethical, biological and social fields, would involve additional risks even more serious than those already run: whoever wishes to gather background information on such problems and feels attracted by the charm of the risks inherent in them, should read Paolo Fenoglio's preface. And then – a wise man has said – the art of boring is to tell it all.

Consequently I'll limit myself to pointing out the main themes of the study.

Unifying factors

The relationship between each of us and the natural structures is ordered by thought. This relationship consists of various unifying factors, to which one should refer in order to understand it better. We are referring to entropy or uncertainty, to the removal of uncertainty or information, to symmetry and the breaking of symmetry, to order, conservation and ambiguity.

The central role of ambiguity

Ambiguity plays a central role among these unifying factors. Where entropy and order, evolution and conservation, symmetry and breaking of sym-

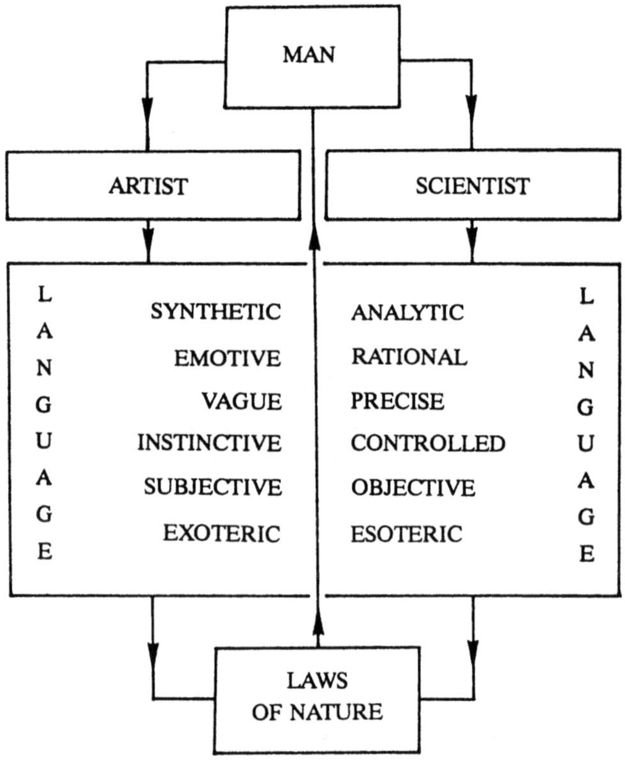

Fig. 38. Language and ambiguity

metry converge, ambiguity appears at the critical point of every choice, even where sensorial stimuli join to become the idea and the idea becomes thought.

Physics and ambiguity

Thus it's no surprise that physicists "by coincidence" sometimes feel prompted to turn to the ambiguous language of "design as visual communication", practically ranking it as a basic language, to analyze and illustrate the structure, properties and behavior of natural systems. Nor, vice versa, should it be surprising that a labyrinthine and unrealizable structure, undecidable and essentially ambiguous, such as those presented repeatedly during this study, should, when perceived, evoke fundamental concepts of quantum physics and spectroscopy, of synergetics and irreversible thermodynamics.

The importance and centrality of the role that ambiguity therefore acquires may prove to be disturbing. But what is the reason behind this destabilizing and, all considered, awkward situation?

"In interiore homine habitat veritas". St. Augustine gives us a cue to the reply to this question. The man-artist and the man-scientist, in searching for beauty and truth, obey the fundamental laws that regulate nature and the mechanisms of its evolution. Both, in revealing their thought, express these laws, each in his own language (Fig. 38). The language of the artist is usually synthetic, vague, instinctive, exoteric and that of the scientist is for the most part analytic, precise, rational, esoteric. But at the point where science and art converge, where truth meets beauty and beauty meets truth, the language becomes analytic and synthetic, precise and vague, rational and instinctive, esoteric and exoteric at the same time. In a word, it becomes *ambiguous*.

Ambiguity at the meeting point of truth and beauty

We then have the form-non-form, the "form-surprise". Labyrinthine and stimulating in its dynamic instability, it becomes the ancestral image of the playful constant factor of taste.

Ambiguity as a permanent cultural value

And ambiguity is elevated to the role of a permanent cultural value.

Appendix I

The Atom According to Quantum Mechanics

A satellite around the Earth, an electron around the nucleus

It would be impossible to try and synthesize the bases of quantum mechanics in a few pages. In this appendix we shall simply explain some key concepts by examining the classic behavior of an artificial satellite moving in the gravitational potential of the Earth and, at the same time, the quantum behavior of the electrons in the atoms.

We shall focus our attention first of all on *energy levels* and subsequently on the atomic wave functions, or stationary *states*, associated to them.

An artificial satellite designed to orbit around the Earth can assume an arbitrary energy value within a continuous interval. In the same way the geometrical features of the orbit can be fixed precisely and without any discontinuity: thus, apparently, natura non facit saltus. The ground state of the artificial satellite, corresponding to minimum energy, is realized when it falls to earth, through a progressive spiralization of the orbit due to a continuous dissipation of energy caused by the friction of the atmosphere. Similarly, for the planet Earth slowed down by the viscous and dissipative alternation of tides, the ground state could be that following the terrifying impact with the Sun – a reassuring prospcct bccause it is far off.

The dynamic behavior of electrons in atoms is quite different.

The red color of a neon light is due to the fact that the atom irradiates (or absorbs) energy according to a spectral distribution that is not continuous but is composed of a discrete and regular group of lines (or colors). Thus one may consider (Bohr's hypothesis) that atomic electrons exist only on *discrete energy levels*. In jumping from a level m with energy E_m to a level n with energy E_n, the electron irradiates or absorbs a gush of electromagnetic radiation having frequency ν_{mn}, or a quantum of light or photon, whose energy

$$h\nu_{mn} = E_m - E_n$$

(h is Planck's constant) is exactly equal to the difference between the energies of levels m and n associated to the states involved in the transition (Fig. 6). The energy levels of the atomic electrons are therefore organized according to a discontinuous distribution of values: natura facit saltus, the behavior of matter is quantized at the atomic level.

Probabilistic meaning of the wave functions

The uncertainty inherent in the probabilistic meaning of the wave or state functions one must use in describing electronic states is of typical quantum origin as well.

Eigenvalues and eigenstates

Every energy level or discrete eigenvalue in which an atom may be found is associated to a state, or better, to a state function. This function is also called eigenfunction or eigenstate of the energy H corresponding to its eigenvalue, and stands for the probablity amplitude of the presence of the electron in any space region around the nucleus. The square modulus of this function gives the probability of the electron's presence in the unitary volume.

The ground state

In the following we shall focus our attention mainly on the minimum energy state. It is called ground state.

Let's refer to the simplest case, the hydrogen atom, formed by a proton and therefore by a single electron. Unlike what happens in the artificial satellite, the electron which is originally in an excited state reaches the ground state after having rid itself of the excess energy through a discontinuous succession of separate irradiating processes. The excited hydrogen atom runs the course of its evolution and reaches its ground state following an exceptionally short time scale.

Again, unlike the artificial satellite, the electron in the ground state of the hydrogen atom hurls itself continuously on the nucleus along orbital segments which branch out from the nucleus in every direction (Fig. 7). On the average, it remains however a certain distance from the nucleus, and is localized in a spherical region whose radius is approximately equal to that of the corresponding classical isoenergetic orbit. In the quantum scheme we must refer to it is meaningless, according to Heisenberg's uncertainty principle, to try to localize the electron's position *exactly* at a certain moment and to measure its momentum *exactly* at the same moment. One cannot expect to make an ideally precise measurement of the position and, *simultaneously*, of the momentum of an atomic electron without inducing a considerable change of state in the system. The precise measurement needed to localize the electron in fact requires specific interaction between the observing instrument, adequately prepared for the measurement, and the electron. It might be necessary for example to verify the electron's position by "lighting it" with a radiation. When the electron, as target, scattered this radiation, it would acquire sufficient energy and momentum for leaving its original state once and for all.

The uncertainty principle

We shall not go into detail here about the modalities of this dynamic interaction of the electron (or any other particle – neutron, proton, atomic nucleus – forming a microscopic object) with the proper macroscopic instrument.

It's enough to note that, qualitatively speaking, the spectroscopic observation of a quantum structure seems to be similar to the perception of ambiguous structures (see Insert on "Symmetry, Information and Ambiguity in Quantum Physics and Design").

Let's return now to the wave function of the electron belonging to the hydrogen atom. As mentioned earlier, the square of this function gives the *probability* of the electron's presence in a certain space region. This probability is a spherically symmetric function of the distance from the nucleus and reaches a maximum around the radius of the classical orbits corresponding to different energy levels E_n (Fig. 7).

Thus the atom, in each of its stationary states, is characterized by a well-defined energy level. But the instantaneous configuration of its electrons – which is spherically symmetric, i.e. isotropic – is essentially uncertain, since we know only the probability with which the electrons can be found at various distances around the nucleus. Whoever wishes to remove this uncertainty, that is *inquire* about the positions of the electrons, must interact with them, as we've said, by means of an instrument for measuring. The very nature of this interaction however is such as to make it foolishly ambitious to design an instrument suitable for completely removing uncertainty without at the same time producing a profound change in the structure under investigation.

The minimum uncertainty with which it is conceivably possible to acquire information regarding the physical quantities pertaining to the state of a quantum structure is fixed quantitatively by the Heisenberg uncertainty principle: the product of the uncertainties of two "conjugated" quantities, such as position x and momentum p, the angle and the angular momentum, time and energy, is no less than Planck's constant, $h = 6.6 \cdot 10^{-27}$ erg s. If we call Δx and Δp the smallest uncertainties associated to a simultaneous measurement of x and p we have:

The Planck's constant

$$\Delta x \cdot \Delta p \gtrsim h.$$

Planck's constant is present in the formal description of every quantum phenomenon. It provides a criterion for distinguishing between large and small, macroscopic and atomistic, classical and quantistic. Its finite numerical value – that is, not vanishing – indicates that absolute space, which in Newtonian mechanics and usual mathematics is conceived as a continuous ensemble of dots, is not similar to physicists' granular space, in which each dot has its own dimension fixed by the precision of the instrument or by the wave-length of the radiation used to measure the distances.

As a matter of fact, according to the uncertainty principle, reducing uncertainty in measuring the position of a particle to ambitiously low values, let's say on the order of $\Delta x \simeq 10^{-40}$ cm, would require the use of a microscope that works with radiations, electromagnetic radiation for instance, whose wavelength is $\lambda \simeq \Delta x$ and therefore of the same order of magnitude 10^{-40} cm; the energy associated to every quantum of this radiation is hc/λ (where $c \simeq 3 \cdot 10^{10}$ cm s^{-1} is the speed of light) equal to 10^{24} erg.

It's a huge value, frighteningly large. Consider that 10^3 erg corresponds to the kinetic energy that a one-gram mass acquires in falling from the height of one centimeter, and that one needs about 10^6 erg for cutting a piece of rock of unitary volume in half, thus creating two new surfaces of a square centimeter each. An energy of 10^{24} erg, like the one associated to the quantum of radiation needed to perform the unrealizable measurement described above, would suffice for creating a fracture surface of 10^{18} cm^2, that is, for cutting the earth into two distinct hemispheres with a single photonic shot.

The numerical value of Planck's constant accounts for the need to use high energy particles for studying the subnuclear world, in which lengths are measured in fractions of *fermi* (1 fermi = 10^{-13} cm). According to what we have said, the numerical value of h contributes to explaining, in a field that is anything but microscopic, the considerable size of particle accelerators and the great outlay of means and energy that characterize physics research in the high energy field.

Getting back to the hydrogen atom, the numerical value of h assures us that the electron will never fall on the proton and remain there as if glued to it, in spite of the great electrostatic attraction one would expect between an electron and a proton at such short distances. Confining the electron on the proton would in fact be the same as localizing it with extreme precision. According to Heisenberg's principle, the uncertainty in the momentum of an electron localized in a region of the order of one fermi, would be quite sufficient for permitting it to escape from the excessive confinement imposed on it.

Uncertainty and stability

The geometrical dimensions, the dynamical features of atoms and their very stability are fixed once the numerical value of h has been fixed. Planck's constant measures the arrogance of uncertainty, so to speak. If Planck's constant were to vanish unexpectedly, the electron of the hydrogen atom, in 10^{-11} sec would fall on the nucleus with exemplary determination, in a spiral of light. A world in which such involutions were imposed by the laws of classical mechanics and electromagnetism would leave no room for uncertainty. Nor for imagination.

Appendix II

The Stark Effect in the Hydrogen Atom

When we apply an electric field to an atom, its original spherical symmetry is broken and a rearrangement of the energy levels and their corresponding electronic stationary states is produced. Spectroscopic measurement of the electronic transitions among various energy levels is concerned with spectral lines which also are perturbed by the action of the electric field. This is the Stark effect.

The hydrogen atom in an electric field

Let's consider the hydrogen atom for instance, and analyze the effect of an external electric field on the energy levels and the stationary states of the electron.

When, because of the atom's spherical symmetry, the field is not applied, the symmetry operations, besides parity, P, are the square modulus of the angular momentum, M^2, and the component M_z of M along a particular axis (classically, e.g. for a planet's orbit, the mass of the planet multiplied by double its areolar velocity).

It follows that the eigenfunctions of H, that is the state functions of this structure, are also eigenfunctions of M^2, M_z and P.

Consequently the search for the state functions and their relative eigenvalues is facilitated by comparing H with the Ω operators corresponding to the transformations mentioned above.

These state functions and their eigenvalues were discussed in Appendix I, emphasizing the ground state energy level, $-Rhc = -13.5\,\mathrm{eV}$ (Fig. 6) and the corresponding eigenstate $1s$. In the following we should keep in mind that immediately above the ground state level is the first excited energy level, $-Rhc/4 = -3.39\,\mathrm{eV}$, and that four eigenfunctions and as many atomic orbitals are associated to this energy eigenvalue: one the $2s$ orbital, gerade and with spherical symmetry; the other three, $2p_z$, $2p_x$ and $2p_y$, ungerade and with axial symmetry around the z, x and y axes respectively (Fig. 17).

What happens to these energy levels and their associated eigenfunctions when we place the hydrogen atom in an electric field E_z, aimed, for example, along direction z?

If the electron is in the ground state, that is, in the spherically symmetric orbital $1s$ (Fig. 7), on the average its potential energy when it is north of the proton increases to the same degree that it decreases when it is south of the proton. Actually, the energy eigenvalue of the ground state remains unchanged.

This does not happen to an electron which is initially in the first excited state, characterized, as we've just said, by the energy level $-Rhc/4 = -3.39\,\text{eV}$. Of the four eigenstates associated to this eigenvalue one need consider only the two orbitals $2s$ and $2p_z$, which have an axial symmetry around the direction in which the electric field E_z is applied. In fact, the other two orbitals $2p_x$ and $2p_y$ for the first excited energy level, have directions orthogonal to direction z of the applied field, and therefore whatever happens along this preferential direction can be ignored (Plate XV).

Symmetry breaking and hybridization

As a consequence of the application of the field the electron's energy varies according to its position, either north of the proton or capsized to the antipodes with respect to it. The parity symmetry is broken and the state function can no longer be forced to remain either *g*erade or *u*ngerade. Thus nothing prevents the two states $2s$ and $2p_z$, both with the same energy and both lacking a component of the angular momentum along the field E_z, from mixing.

In their linear combination due to the presence of field E_z, the two original eigenfunctions $2s$ and $2p_z$, associated to the energy level $-Rhc/4$, give rise to a pair of hybrid atomic orbitals, as shown in Fig. 17. The hybrid atomic orbital $(2s + 2p_z)/\sqrt{2}$, whose main lobe points in the same direction as the field, describes an electron on the average shifted to the north with respect to the proton; the other hybrid atomic orbital, $(2s - 2p_z)/\sqrt{2}$, whose lobe points in the direction opposite to that of the field, describes an electron on the average shifted to the south with respect to the proton.

The energy of an electron described by the first of these two hybridized eigenfunctions is obviously greater than that of the one described by the second eigenfunction. The original energy level $-Rhc/4$, where the eigenvalues of $2s$ and $2p_z$ converge, is split; in the hydrogen atom's spectrum satellite lines appear adjoining those which in Fig. 6 connect the level characterized by quantum number $n = 2$ with other levels.

The Stark spectrum

Concluding, the application of an electric field E_z has a two-fold effect. It reduces the original spherical symmetry of the atom, and certain atomic orbitals perturbed by the field, combining with each other, modify their shape choosing the direction of the applied field as the direction of symmetry: the reduction of symmetry allows eigenfunctions ($2s$ and $2p_z$ for example) pertaining to the same energy level to mix. Furthermore, the application of E_z modifies the arrangement and/or the spacing of the energy levels relative to those orbitals. These levels split, causing a change in the spectroscopic and dynamic behavior of the atom. New lines appear in its spectrum: the Stark lines.

Appendix III

Maxwell's Demon: Physical Entropy and Informational Entropy

This appendix deals with the experimental evidence of the equivalence between physical entropy and informational entropy, and explains the physical meaning of this equivalence according to Brillouin.

Physical entropy, informational entropy

Let us refer to the order $\rightleftarrows$ disorder transformation in the binary alloy CuZn (see Chapter 4, §2). We said that the measure of the order in the alloy depends on the balance between two competing factors:

Order and disorder in a binary alloy

- the difference between attractive potential energy among unlike atoms and attractive potential energy among like atoms;
- the product of the Boltzmann constant by the equilibrium temperature of the alloy, K_BT.

At temperatures slightly higher than the critical temperature T_c we can ignore the first of these two competing factors: that is, the second one, which causes disorder, prevails. The mechanism that makes it possible to achieve a generic microscopic configuration of the alloy can therefore be described in this way.

Statistical interpretation of disorder

Suppose we place $N/2$ atoms of zinc and $N/2$ atoms of copper in an urn and make successive drawings at random, as in a lottery. As the atoms are drawn, they are sown in the lattice sites according – for example – to a boustrophedonic scan of possible atomic positions, pre-arranged so as to fill all the lattice sites gradually. At first sight, the number of different configurations one may obtain after N drawings can be calculated in the following way. The first drawing can give N results, the second one $N-1$ and so on until the last drawing which gives the obligatory result of the last atom left in the urn. Hence, there would be $N(N-1)\cdot\ldots\cdot3\cdot2\cdot1 \equiv N!$ configurations or complexions that one may, a priori, obtain.

Actually, since one should not draw distinctions among atoms of the same species, all the complexions obtained from an initial configuration given by permuting – that is, mixing up – like atoms, are equivalent. The number of possible permutations of $N/2$ copper atoms and $N/2$ zinc atoms is $(N/2!)^2$ so the different configurations are really

$$g = \frac{N!}{(N/2!)^2}.$$

This number, as one senses observing Fig. 39, rapidly becomes quite large with increasing N.

At this point we may wonder by how much the number g just calculated of distinct configurations differs from the one obtainable with the following, simpler procedure.

Let's put in the urn an equal number of zinc and copper atoms, in all much larger than N; as before, let's go on to making successive blind drawings and then placing the atoms in the lattice sites. The result of the first extraction corresponds to a binary choice between two possibilities – either copper or zinc. In every successive drawing there are still two possible results. The number m of different configurations one can obtain with this procedure following N binary choices is obviously equal to

$$m = 2^N.$$

The Tartaglia triangle

As Fig. 39 indicates, g and m are connected by the fact that in the Tartaglia triangle, g is the largest of the binomial coefficients summed up in the binomial expansion $m \equiv (1+1)^N$. While the ratio between g and m decreases with increasing N, the ratio between the logarithms tends quickly to one, so that the latter tend to coincide.

The logarithm in base 2 of g, $\log_2 g$, denotes the number of binary choices which generate all the distinguishable configurations, compatible with the one-to-one stechiometric ratio between copper and zinc. Apart from a numerical factor we shall deal with shortly (see also Chapter 2, §4) the $\log_2 g$ is the same as the entropy of the alloy. $\log_2 m$ (a number which in reality coincides with $\log_2 g$) that is, N itself, signifies the number of binary choices that generate all possible configurations for the copper-zinc system – including the improbable ones for pure copper and pure zinc. The fact that $\log_2 g$ and $\log_2 m$ tend to coincide as N increases allows us to use $\log_2 m$, that is, the second extraction process, to calculate the entropy of the alloy.

Statistical interpretation of entropy

If we adopt this procedure, the statistical meaning of entropy in terms of disorder then becomes intuitive; moreover, the reason why the aggregates of atomic or molecular systems tend to assume the most disordered configuration – it being more probable – becomes plausible. In fact, if the cause which tends to establish the order (in this case, the difference between the interaction energies of like atoms and unlike atoms), is really negligible, one can be practically sure that, for example, the crystals of copper and zinc will not become segregated inside the alloy. It's rather like roulette, where there may be ordered sequences of consecutive blacks and reds or of blacks and reds that alternate in an apparently systematic way, but one can almost exclude that these sequences continue for hundreds of turns.

As we've said however, unlike what happens for the black and red numbers that are uncorrelated in roulette, the copper and zinc atoms in the alloy attract each other into adjacent lattice sites more than like atoms do. What has been said regarding numbers g and m of the possible configurations is valid, strictly speaking, only where temperatures are higher than the critical or Curie temperature T_c, that is, where thermal energy $K_B T_c$ which induces

The critical temperature

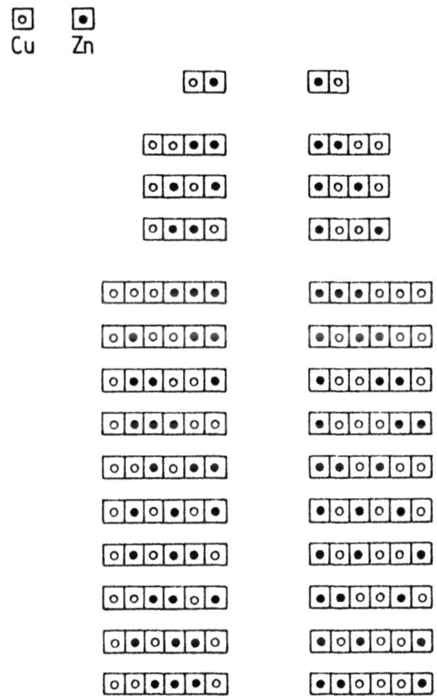

N	$m = 2^N$	g	$\log_2 m$	$\log_2 g$
0	1	**1**	0	0
1	2	1 1	1	
2	4	1 **2** 1	2	1
3	8	1 3 3 1	3	
4	16	1 4 **6** 4 1	4	2.58
5	32	1 5 10 10 5 1		
6	64	1 6 15 **20** 15 6 1	6	4.32
8	256	**70**	8	6.13
10	1024	**252**	10	7.98
10^2	1.27×10^{30}	$\mathbf{1.01 \times 10^{29}}$	10^2	96.3
10^3	1.07×10^{301}	$\mathbf{2.70 \times 10^{299}}$	10^3	994.7
10^4	1.995×10^{3010}	$\mathbf{1.552 \times 10^{3008}}$	10^4	9993.03

Fig. 39. Configurations or possible ways of arranging copper and zinc atoms in the cubic structure of the alloy CuZn

the stochastic diffusive processes that generate disorder, competes effectively with the interaction energy among atoms, which generates order. It follows that we can be almost certain of being unable to obtain an ordered alloy in equilibrium at a higher temperature than T_c. Vice versa, if a crystal of CuZn is cooled down slowly enough to give the atoms time to diffuse and occupy the sites belonging to them because of the differences between interatomic potential energies, an ordered structure is obtained at the end.

In order to clarify the correlation between entropy and disorder, we shall now describe in a thermodynamical framework the statistical aspects already introduced to characterize the total order (or information) in the alloy.

The minimum entropy change conveyed by a bit

Because of its importance, the result of this thermodynamical approach has already been anticipated in §2, ¶4: physical entropy and informational entropy are equivalent. The acquiring of an information bit referring to a given structure, or its intake by the structure itself corresponds to an entropy change not less than $K_B \log_e 2 \equiv 0.97 \cdot 10^{-23}\,\mathrm{J\,K^{-1} mole^{-1}}$.

An amount Q of heat transferred to a sample of ordered alloy which is initially at temperature T_0, much lower than critical temperature T_c, induces a temperature rise in the sample. This rise in temperature is accompanied by an increase in the amplitude of the thermal motion of the atoms around their official positions in the crystal lattice, and a reshuffling of the copper and zinc atoms. In other words, the introduction of an amount Q of heat at a given temperature T produces increased disorder and entropy ($\Delta S = Q/T$). This is partially greater *vibrational disorder* of the atomic thermal motion, and partially greater *configurational disorder* in the distribution of the atoms. The vibrational contribution – ΔS_v – to the total entropy may be separated from the configurational contribution – ΔS_c – by means of calorimetric measurements. (These contributions are proportional to the increase per cent $\Delta T/T$ of the temperature, and to the vibrational specific heat $C_\mathrm{v}(T)$ and the configurational specific heat $C_\mathrm{c}(T)$, respectively).

Configurational entropy of an alloy

For instance, Moser's experimental data (1936) on the specific heat of the alloy CuZn at various temperatures, give a value of $4.9\,\mathrm{J\,K^{-1}\,mole^{-1}}$ for the increase in the configurational entropy of the alloy in the transition from an ordered low-temperature phase to the disordered phase just below the melting point. Fig. 40.

This numerical value for the configurational entropy of a disordered binary alloy, obtained on the basis of experimental data, should be compared with the configurational entropy $S_\mathrm{c}^{(i)}(T)$ associated to one mole of an ideally disordered alloy, that is, to a sample of alloy whose mass, given in grams, is equal to the average atomic weight of Cu and Zn. Such a sample contains a number of lattice sites equal to Avogadro's number, $N = 6.02 \cdot 10^{23}\,\mathrm{mole^{-1}}$.

For this system the configurational entropy may be written as:

$$S_\mathrm{c}^{(i)}(T) = K_B \log_e W$$

where, as you will recall, $W = 2^N$ is the number of complexions corresponding

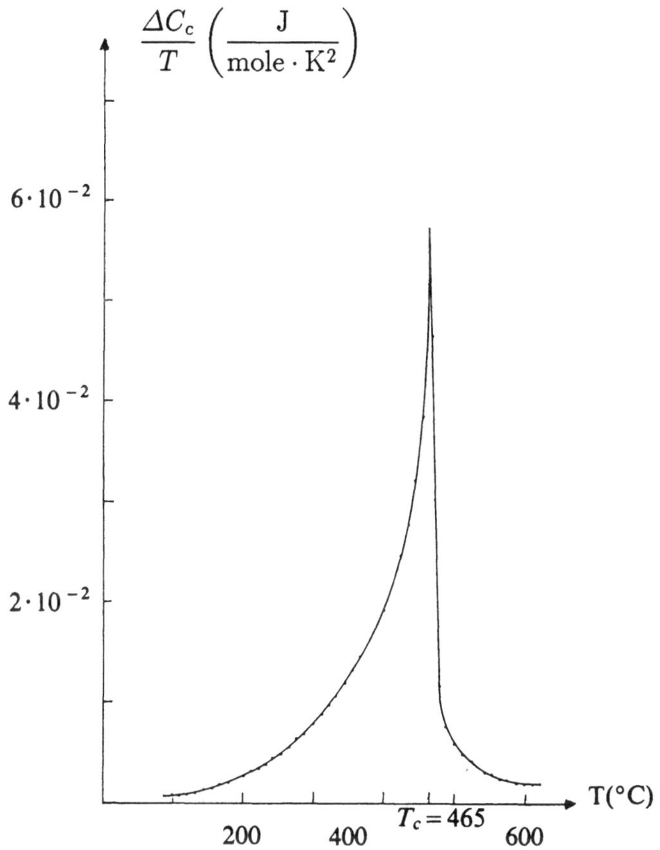

Fig. 40. Behavior of the increase of the entropy of mixing as a function of temperature, around the critical temperature for the order-disorder transition in a CuZn alloy (Moser, 1936). The specific heat, which is the heat needed to increase by one degree the temperature of one mole of the alloy, is given by two terms: the vibrational specific heat and the configurational specific heat. The vibrational specific heat, $C_V(T)$, increases slowly and monotonically with the temperature; its product is the amplitude enhancement of the thermal motion of the atoms around their crystalline lattice sites. As the temperature is varied around the order-disorder transition temperature T_c, the configurational heat, $C_c(T)$, presents a maximum; its product is the reshuffling of the atoms (Fig. 19b), which were arranged in an orderly pattern on the lattice sites at low temperature (Fig. 19a). The configurational specific heat increases until the reshuffling of the species is complete, and then it drops abruptly when the critical temperature of the order-disorder transition, T_c, has been exceeded. The increase in the entropy of mixing, $\Delta S_c = C_c(T)\Delta T/T$ relative to each temperature interval ΔT, is calculated starting from the configurational specific heat absorbed at each temperature T by the reshuffling process. The value of the configurational entropy pertaining to the order-disorder transformation, equal to $4.9\,\mathrm{J\,K^{-1}mole^{-1}}$, is obtained by adding up all the increases, ΔS_c, contributing to the maximum around T_c. This value is interpreted in the text in terms of informational entropy associated to a number of binary choices (*bit*: copper or zinc) equal to the number of lattice sites: a number which, per mole of alloy, is equal to Avogadro's number. In other words, $4.9\,\mathrm{J\,K^{-1}mole^{-1}}$ represents the uncertainty created in the previously ordered alloy by the introduction of 1 bit of entropy per lattice site

to the theoretically disordered macroscopic state of the alloy. Since $\log_e W \equiv \log_e 2 \cdot \log_2 W$, and furthermore $\log_e 2 = 0.699$ and $\log_2 2^N = N$, one has:

$$S_c^{(i)}(T) = 0.699\, N K_B = 5.76\,\mathrm{J\,K^{-1} mole^{-1}}$$

This is the entropy associated to a number of uncorrelated binary choices equal to Avogadro's number N, when an entropic content corresponding to the product of Boltzmann's constant, $K_B = 1.38 \cdot 10^{-23}\,\mathrm{J\,K^{-1} mole^{-1}}$, times the natural logarithm of two, $\log_e 2 = 0.699$, is attributed to each binary choice (bit). In this context the following equivalence, expressed by L. Brillouin, between physical entropy and uncertainty, or informational entropy, holds.

The value of a bit of information in terms of physical entropy

$$1 \text{ bit of information} = K_B \log_e 2 = 0.97 \cdot 10^{-23}\,\mathrm{J\,K^{-1} mole^{-1}}.$$

The small difference (on the order of 15%) between the experimental value $4.9\,\mathrm{J\,K^{-1}\,mole^{-1}}$ and the ideal computed value for the configurational entropy of a disordered binary alloy may be atributed partially to the uncertainty arising from the difficulty of distinguishing between vibrational and configurational contributions to the total specific heat in the experimental results reported by Moser. It is also due in part to the fact that the experimental value, $4.9\,\mathrm{J\,K^{-1} mole^{-1}}$, refers to an alloy in which a short range order persists even at temperatures T higher than the critical temperature T_c.

To sum up, in passing from a temperature T_0 much lower than the critical temperature T_c to a temperature above T_c, thermal energy, which generates configurational disorder in an alloy, prevails over interatomic potential energy, which favors order. Meanwhile, entropy increases. The number of all possible binary choices which may be used to realize the generic macroscopic configuration of the alloy increases with entropy until it coincides, in theory, with the total number N of the atoms; in the completely disordered alloy, each site corresponds to a binary choice. Moreover, at each temperature the number of choices – and therefore the configurational disorder – takes at equilibrium the highest value compatible with the tendency of the potential energy to reach its minimum value.

The observations discussed above may be extended to all physical-chemical systems. For example, regarding an ordered solid which melts on absorbing latent heat of fusion L at the melting temperature T_m, we might say that the increase of entropy L/T_m which accompanies this (first order) phase transition implies judgemental uncertainty concerning $L/(T_m K_B \log_e 2)$ binary choices.

The above has enabled us to establish an equivalence between informational entropy, computed on the basis of combinatory analysis of the number of possible binary choices, and physical entropy, measured starting from the behavior of the specific heat of the copper-zinc alloy as a function of temperature. Given the importance of this subject, the rest of the appendix will be dedicated to showing how Brillouin clarified the physical basis of this equivalence by critically analyzing the measuring process (i.e., the acquiring of information or the removal of entropy), solving at the same time the enigma

The cost of the information

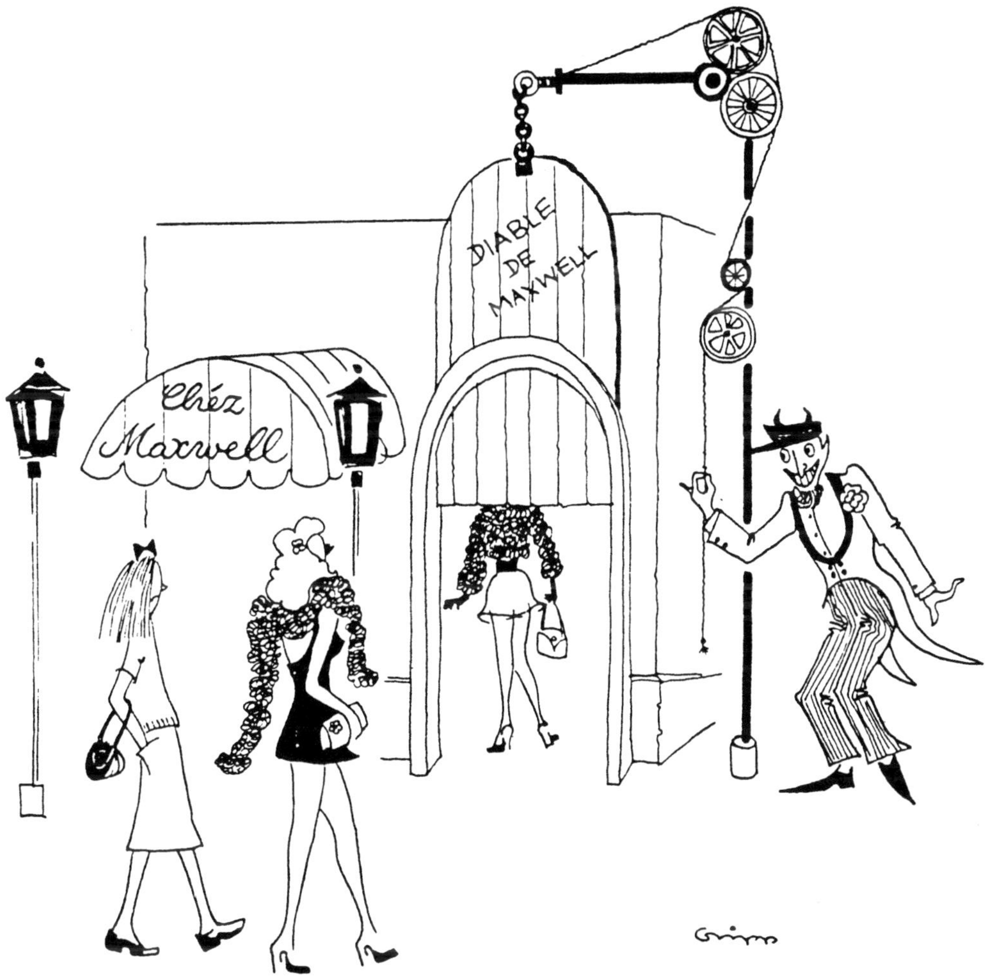

Plate XVI. Anneliese Gripp. The Maxwell demon at work

presented by the most stimulating of all the paradoxes devised to invalidate the second principle of thermodynamics, namely Maxwell's paradox. What's it about?

The Maxwell's paradox

An isolated system is formed of a demon and two ampullae filled with gas (e.g. helium) and connected by a pipe which contains a trap in its middle section (Fig. 41). The demon triggers the trap without energy loss every time an atom faster than the average, coming from ampulla A and directed towards B, passes through. In this way, the demon concentrates the faster atoms, with higher than average kinetic energy, in ampulla B, while relegating the slower atoms, with below average kinetic energy, to the other ampulla. Maxwell, who together with Boltzmann, developed among other things the kinetic theory of gases, was well aware that the temperature T of a gas measures the average kinetic energy, $\overline{E}$, of its molecules or – as in our example – of its atoms.

The Maxwell's demon

Therefore he knew the equation that expresses the principle of equipartition of energy:

$$\overline{E} = 3K_BT/2.$$

A diabolical performance

The demon, working inside the system formed of a gas whose atoms have a mean kinetic energy $\overline{E}$ and were originally in thermal equilibrium at temperature T, thus manages to concentrate the atoms having energy greater than $\overline{E}$ in an ampulla that reaches temperature T_1 above T, and the atoms having energy less than $\overline{E}$ in the other ampulla, which reaches temperature T_2 less than T (Fig. 41). The demon therefore carries out a transformation whose only result is to trigger two heat sources at different temperatures, T_1 and T_2, starting from a single heat source at temperature T. What manna for the electric companies, which would hasten to install a thermal machine capable of supplying electric energy or work between these two sources, kept permanently at two different temperatures by the tireless demon: with the sole result of having made, and of being able to use gratuitously, a transformation in which the thermal energy originally available at a single temperature is converted into work. This result is in contrast with the second principle of thermodynamics. A diabolical result. Where is the flaw in the reasoning?

Not in having hypothesized that the guillotine in the trap may be worked without expending energy. To an idealized world, perhaps even without friction, where the direction in which time flows is indifferent (a spatialized time, reduced to the level of a parameter associated to motion, crystalized in a succession of equivalent instants), Galileo had accustomed even those who in using their time took their inspiration from Seneca[1]: ridding reality of elements believed to be temporarily secondary or non-essential is a methodological premise to all physical research on natural systems. The weak point of the argument is another.

Maxwell's demon is exorcised: information leads to order, but information costs

To acquire the necessary information, that is, to distinguish swift atoms from slow ones, the demon must at least be able to observe them. The demon, equipped with a light, must therefore illuminate them and analyze the radiation which – diffused by the atoms – strikes its eye. To be clearly perceived in the thermal bath at temperature T, this radiation must be composed of $h\nu$ energy photons greater than K_BT. Each photon absorbed thus increases the demon's entropy by a quantity ΔS_d, comparable to Boltzmann's constant:

$$\Delta S_d = h\nu/T > K_B.$$

[1] From Lucius Annaeus Seneca, Letters to Lucilius, Book I, Letter I: Now then, dear Lucilius, do as you say; take advantage of all the time you have. You'll be less the slave of tomorrow, if you've become the master of today. Life passes while we put off our commitments. Everything, O Lucilius, depends on others; only time is ours. Nature has given this one, excessively fleeting good into our keeping, but we let the first one who passes take it away. And man is so that, when he acquires goods that are worthless, or at any rate compensatable, he agrees that they be put on his bill; but no one who has caused others to lose time, thinks himself to be in debt, while this is the only good that man cannot repay, not for all the good will in the world.

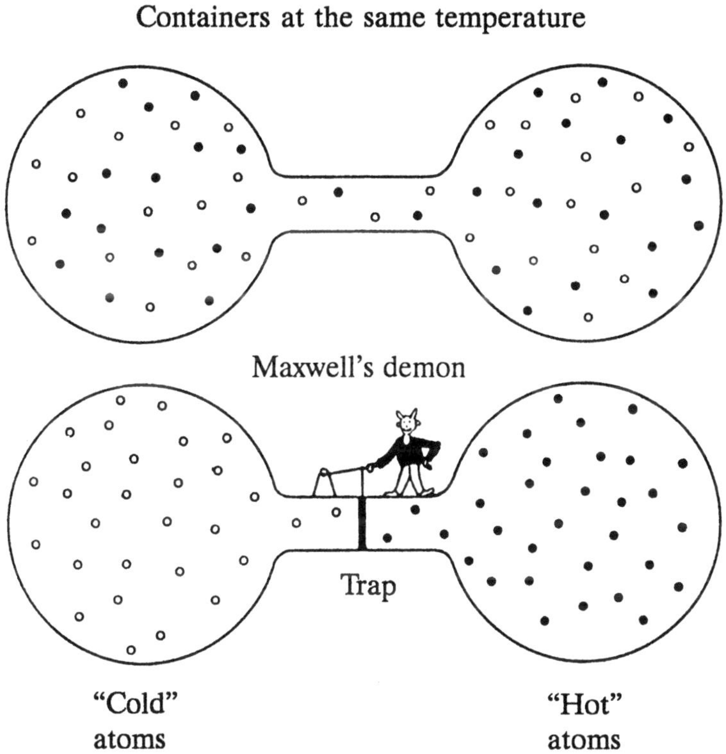

Fig. 41. In his Theory of Heat, James C. Maxwell (1871) wrote: "One of the most confirmed facts in thermodynamics is the impossibility, for a system closed to volume changes and hcat passages, and in which temperature and pressure are the same at every point, of producing any inequality of temperature or pressure without spending energy. This is the second law of thermodynamics, and it is certainly true; as long as we can deal with bodies on a macroscopic scale, we cannot perceive and handle the individual molecules of which they are composed. But if we imagine a being whose faculties are so acute that he can follow every molecule in its course, this being, whose characteristics are essentially finite like our own, would be capable of doing what is impossible for us at present. We have seen that the molecules in a container full of air at a uniform temperature move at speeds that are not uniform at all, although the average speed of each large sub-group is practically uniform; suppose that this container is divided into two parts, A and B, by a wall containing a small hole, and that a being, who can see the individual molecules, opens and closes the hole so as to allow only the fastest molecules from A to B and the slowest from B to A, to pass. Without working, he would increase the temperature of B and lower that of A, contrary to the second law of thermodynamics". This is Maxwell's famous paradox. Actually, "you can't get something for nothing, not even an observation": getting information costs. And even if one takes advantage of the information acquired by re-ordering the gas atoms efficiently on the basis of their energy content, the energy needed to obtain the information is greater than the work that may be regained from the system re-organized in this way. Moreover, to acquire information on the molecules' energy, the demon produces more uncertainty in the isolated system he belongs to than that removed by informing himself. In other words, information generates a greater degree of disinformation than that corresponding to the uncertainty removed. (From A. Lehninger, op. cit.)

The demon immediately stores up the entropy increase ΔS_d that has struck its eye, mentally converting part of it into information on the rest of the isolated system it belongs to. The information acquired enables it to separate the gas atoms into the two ampullae, according to their energy content. This reordering implies a decrease ΔS_g in the physical entropy of the gas. The decrease, which results from the demon's having acquired a bit of information (the atom's energy is more or less than $K_B T$), corresponds to $\Delta S_g \simeq -K_B \log_e 2$. We thus find

$$\Delta S_d + \Delta S_g > 0,$$

a consequence of the second principle of thermodynamics. At the same time, Maxwell's demon is exorcised.

Acquiring information implies an uncertainty production larger than the removed uncertainty

It's true in fact that the demon, by reducing the physical entropy of the rest of the isolated system it belongs to, enables the gas in the ampullae to produce work. But it's also true that the price the devil pays for acquiring the information that permits it to operate usefully is extremely burdensome: even eliminating waste, the devil must collect physical entropy of not less than $K_B \log_e 2$ in its eye to acquire a bit of information and reduce the informational entropy of the brain by $K_B \log_e 2$. Acquiring information, that is, removing uncertainty or informational entropy in part of an isolated system is possible only if, at the same time, the entire system's physical entropy increases on the whole by at least as much as the informational entropy decreases locally.

Regarding the energy aspect on the other hand, it's true that every fast molecule trapped by the devil in the warmer ampulla can make a contribution, on the order of $K_B T \log_e 2$, to the work produced by a thermal machine that functions by using the two ampullae as a heat source. But it's also true that the price the devil pays in energy for acquiring the information that enables the system to produce work is no less than $K_B T \log_e 2$, the total amount of the work produced.

Thus Maxwell's paradox does not invalidate the second principle of thermodynamics: but one must admit a substantial equivalence between physical entropy and informational entropy. Removing uncertainty or informational entropy costs: at the thermodynamic limit, to acquire a bit of information, in entropic terms one needs $K_B \log_e 2 = 0.97 \cdot 10^{-23}\,\mathrm{J\,K^{-1}}$, and in terms of energy, at room temperature, $K_B T \log_e 2 \simeq 10^{-27}$ KWh.

And the demon? Brillouin has exorcised it, but we won't throw it away: with its managerial abilities re-evaluated, but inflexible in its absurd zeal, it isn't as bad as it seems. We too, for storing and elaborating information as we prefer, and for establishing and perpetuating order – especially genetic order – through information, need a torch – as does the devil – that is, a source of entropy and energy. Whether we live with nuclear energy or with oil, or whether an ever more complex world actually manages to go back to "living with the sun", the most ancient fount of energy, the chosen source, at its origin is, in the last analysis, external. In converting part of the energy taken

Energy as a source of order

from the source into information, the demon, and the living organisms, by generating order, improve the quality of the energy that is made available to them: in the "mill of life", we can recognize,

> "a singularly useless machine, which with an earnestness strangely out of proportion to the absurdity of the function, with care and dedication devours all the energy obtained from the source; all its work goes into feeding and maintaining itself, so that nothing is left over for any further purpose. But it does one very important thing: it improves itself, if we may use this term to describe those progressive changes in composition and construction that make up the evolution of the system". (Lotka, 1956).

Glossary

Ambiguity

The coexistence, at a critical point, of two mutually incompatible aspects of the same reality.

The existence of ambiguity, regarding a structure, assumes the existence of a symmetry, which may be broken however by an external action (giving a control parameter increasing values).

Let's consider for instance the contiguous cubic moduli in Plate IV, and hypothesize an analogy between the sensorial process leading to the dynamic perception of an ambiguous structure and the physical process leading to a second order phase transition. As long as the structure appears to be center-symmetric, the inner wall, by its structure, belongs to both the modulus on the left and the one on the right, or rather it doesn't belong to both the modulus on the left and the one on the right. But this wall seems to move to the right or the left in the critical moment in which the inversion or parity symmetry is broken by visual thinking. And, contemporaneously, ambiguity ceases to exist.

Angular Momentum and Rotations

In classical mechanics the angular momentum of one point (a satellite, for instance) around another point (the Earth, for instance) is double the product of the satellite's mass by the areolar velocity with which the satellite covers its orbit. Observations similar to those made for momentum and translations are valid also for the angular momentum and the rotation angles.

Catalectic Trochaic Tetrameter

A verse formed of eight trochees, the last one truncated by the absence of the final short vowel. It lends itself to a markedly repetitive metric scansion.

Control Parameter

Measures how far away a system is from the critical point of a phase transition or the bifurcation that leads to a dynamic instability. The control parameter may be associated both to phase transitions (second order) between states of thermodynamic equilibrium in a system, and to dynamic instabilities between a state of equilibrium and a dissipative structure (bifurcation of the thermodynamic branch) or between dissipative structures in cascade. This may usually be expressed in terms of difference between "gain" and "loss"; that is, between an external destabilizing action and the system's dissipative reaction, which tends to maintain it in the status quo. When the destabilizing action wins out, the critical fluctuation associated to one of the system's collective variables, instead of regressing, emerges as the order parameter, and the system organizes itself according to a new structure. Two examples of control parameter:

- the temperature decrease in a paramagnetic-ferromagnetic phase transformation
- the temperature difference ΔT among the bases of the Rayleigh-Bénard capsule in the dynamic instability that leads to the convective regime (see Dissipative Structure).

Dissipative Structure

An expression introduced by I. Prigogine to indicate a structure according to which a system may organize itself, when it is removed from thermodynamic equilibrium: in such a structure a dynamic order is maintained at the cost of energy injected, at the necessary rate and in sufficient quantity, from outside the system, and then dissipated within the structure.

For instance, the establishing of an ordered, convective movement of fluid rolls in a Rayleigh-Bénard capsule among whose bases a temperature difference ΔT, which exceeds a critical threshold $(\Delta T)^*$, is maintained by spending thermal energy administered from below, is accompanied by large gradients of the temperature field within the fluid. Consequently, the dissipation of energy due to these gradients is well above that which takes place below the critical threshold, that is for ΔT less than $(\Delta T)^*$.

Eigenvalues and Constants of Motion

According to quantum mechanics – unlike what occurs in classical mechanics – the possible results of measurement of an observable may define a discrete and not necessarily continuous set of values (see Fig. 6).

These possible results are the eigenvalues of the operator associated to the observable, and constitute the constants of motion.

Entropy

From the Greek ἐντροπή, evolution.

A definition of this important function of the state of a thermodynamic system is given on page 72. As a function of the thermodynamic state, entropy usually varies when the system undergoes a transformation from one state to another. The lower the entropy of a system in a state of fixed entropy, the higher the system's evolutionary potential toward states of greater entropy or disorder, should the restraint be removed. Thus entropy represents a kind of potential for the system's future evolution.

Should the system undergo a cyclic transformation and return to its original state, its entropy does not change. If an isolated system undergoes an irreversible transformation (think of a spontaneous evolution of the system, for example), the isolated system's entropy increases.

At any rate, the production of entropy inside a system, even a non-isolated system, is never negative. A positive production of entropy implies dissipation, irreversibility and eventually the creation of new structures.

In a system at equilibrium, entropy is a measure of the disorder and uncertainty present in a system, that is, of the absence of correlations among the structural moduli making up the system. There are many forms of disorder which may contribute to entropy: configurational, thermal, magnetic, ...

The formal expression of entropy coincides, except for the product of $\log_2 e$ by Boltzmann's constant, with the expression of informational entropy given under that entry.

Information

The removal of uncertainty or informational entropy according to L. Brillouin; informational entropy according to C. E. Shannon. Brillouin's information coincides substantially with Shannon's redundance (see also Informational Entropy).

Information Bit (see Informational Entropy)

Informational Entropy

It measures the uncertainty inherent a priori in the choice of alphabetical symbols. Calling p_i the appearance probability of an alphabetical symbol, informational entropy S is expressed as

$$S = -\sum_i p_i \log_2 p_i,$$

where $\log_2$ (base 2 logarithm) is the exponent to which the base, 2, must be raised to obtain p_i.

For example, in an alphabet of only two letters each having the same probability, equal to 1/2, of appearing in a message,

$$S = -(1/2)\log_2(1/2) - (1/2)\log_2(1/2);$$

since the exponent to which it's necessary to raise 2 to obtain $1/2 \equiv 2^{-1}$ is -1,

$$S = -(1/2) \cdot (-1) - (1/2) \cdot (-1) = 1.$$

In this case we say that the informational entropy pertaining to the generic alphabetical symbol is equal to one. This entropy is measured in bits, and S is equal to 1 bit/symbol: the unit of informational entropy, or bit, is associated to the basic alternative – heads or tails for a coin, black or red in roulette – between two equiprobable possibilities.

In the ordinary alphabet, if the twenty-six letters plus the blank space were equiprobable (with probability equal to $p_i = 1/27$), informational entropy $S_{\max}$ would be equal to

$$S_{\max} = -\sum_{i=1}^{27} p_i \log_2 p_i = 27\left(-\frac{1}{27}\log_2\frac{1}{27}\right) = \log_2 27 = 4.7\,\text{bit/symbol}$$

Actually, the game of identifying the text of a message, described on page 85, makes it possible to ascertain that the number of bits associated to each symbol in a message is more or less two: the informational entropy of a message is approximately 2 bit/symbol.

The syntactic organization of a language, an expression of both tradition and culture, thus makes redundant – in Shannon's sense – about 2.7 (= 4.7−2) bits for each symbol (see following scheme).

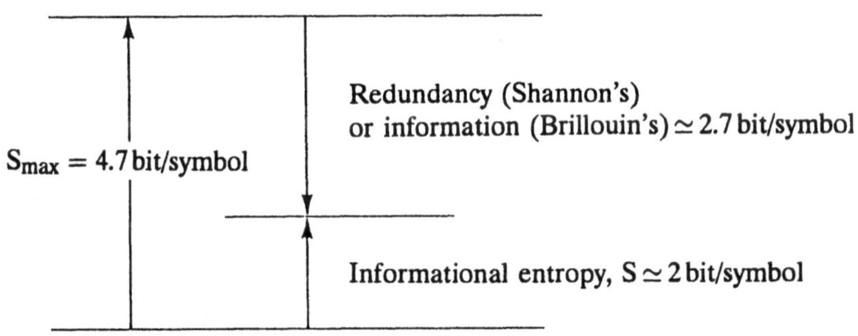

Momentum and Translation

In classical mechanics the momentum is the product of the mass of a system by the velocity of its center of mass. In quantum mechanics the momentum is an observable to which an operator is associated. The possible experimental values of the observable momentum are the eigenvalues of the momentum operator. In an isolated system having translational symmetry – a crystal, for example – the eigenvalues of the momentum are constant and are called constants of motion. Vice versa, a translational symmetry corresponds to the constant values of the momentum.

The eigenvalues of the momentum are inversely proportional to the lattice steps associated to the translation. The momentum operator "generates" the translation. Constant and "small" ("large") eigenvalues of the momentum correspond biunivocally to "high" ("small") lattice steps, which are also constant.

Observable

A physical quantity susceptible of being measured. Energy, momentum, angular momentum of a quantum system are some typical observables.

Operator and Energy Operator

A procedure which changes one function into another function.

In quantum mechanics, an observable is associated to each operator.

For example, the energy operator is associated to the energy observable. The energy operator transforms every function of the variables (position, momentum, ...) on which the state of a quantum system (an atom, a molecule, a crystal) depends, into *another function*. However, if the function that the energy operator is applied to coincides with a state function or eigenfunction of the system, the energy operator changes this function into *itself*: or, better, into the product of the state function by a constant quantity called eigenvalue of the energy associated to the eigenfunction.

Order

The correlation among structural moduli. It is the opposite of disorder, and is measured by the decrease of entropy, or neg-entropy.

Order Parameter

In a thermodynamic system composed of N material points, there are $3N$ degrees of freedom: that is, there are as many degrees of freedom as there are possible choices of the three spatial coordinates associated to each of the N material points. Describing the motion of each of the N material points, an undertaking that was humanly impossible before the advent of computers, is still difficult today when N goes beyond a few thousand, but it is often superfluous. One usually resorts then to macroscopic variables of a collective nature. In systems taken out of thermodynamic equilibrium beyond a certain threshold of the control parameter, a critical fluctuation pertaining to one of the above-mentioned variables may be destabilized instead of regressing. This then ends by *enslaving* the entire system, and appears to be the only macroscopic degree of freedom that is really significant and representative of the structure or of the system's behavior. This accounts for its being called the order parameter. Think for example of the amplitude of an electromagnetic field's basic mode in a laser cavity, or of the cadenced intensity of rhythmic applause at the end of a good concert.

Phase Transition and Dynamic Instabilities

The passing of a system from one structure whose elements are characterized by certain geometric and/or dynamic correlations, to another structure characterized by different geometric and/or dynamic correlations among the moduli that compose it.

A phase transition is "first order" if at the critical point the two structures in equilibrium with each other are different (think of the ice-water system); it is "second order" if they coincide (think of the brass-zinc alloy for which, at the point critical to equilibrium, the ordered and disordered states of the alloy's structure coincide). There is a close analogy between second order equilibrium phase transitions and a wide class of non-equilibrium dynamic instabilities, for which the order parameter which emerges in the dissipative structure grows with continuity starting from zero.

Symmetry and the Breaking of Symmetry

From the Greek συμ·μετρία, what was originally meant for sym·metry was a com·mensuration of relationships among the elements or moduli of a structure: according to Pythagoras' canon, for a stringed instrument to produce the musical note corresponding to the octave, or to the fifth or fourth superior of the vacant chord, the chord must be blocked at its mid-point, that is at half (1/2) of its length, or at 2/3 or 3/4 of its length, respectively. Musical notes which are in harmony with the note of the vacant chord correspond to the com·mensurate relationships among lengths, expressed by "small" integer numbers. Hence symmetry is also a synonym for harmony.

In current scientific language, the accepted meaning includes a com·mensuration of relationships among structural moduli that is so extreme as to be actually immensurable: that is, any difference between the designated structure and the structure obtained by operating a symmetry transformation on the designated structure, is indiscernible: think of a glass rotating around its axis, for example. Thus a transformation is called a symmetry transformation if it leaves the structure unchanged: symmetry is *invariance* as the result of a transformation; it is *indiscernibility* of the change produced by a transformation. Symmetry assumes the impossibility of measuring or perceiving the alteration of basic quantities, characteristic of the structure. In this sense, symmetry is more like entropy or uncertainty than order or correlation.

A breaking of symmetry involves the establishing of correlations, that is of order, among structural moduli originally having a relationship of indifference to each other. Symmetry assumes the indiscernibility of transformations; the breaking of symmetry implies an awareness of transformations and is therefore contemporaneous to the production of information.

For instance, when parity symmetry is broken in an ambiguous figure, indifference or rather a balanced relationship between right and left is replaced by a perceptible or measureable differentiation between right and left (see also

Ambiguity): information is obtained (right or left) contemporaneously with the breaking of symmetry.

Synergetics

A neologism coined by H. Haken (from the Greek *συν·ἔργον*, col · laboration) to mean the joint action of elements of a structure which, under the influence and beyond a critical threshold of a control parameter, organize themselves in a dissipative structure characterized by an order functional in space and/or time (see Insert p. 7).

Spectroscopy

Analysis by absorption, diffusion (and possibly by emission) of appropriate radiations (for example, visible radiation, infrared, ultraviolet, X-ray, ... , or electronic, neutronic ... radiation), of the temporal and/or spatial frequency typical of a structure (atomic, molecular, crystalline, ...), that is, of the elementary excitations (electromagnetic radiation quanta or photons, phonons, magnons, ...) for which the structure itself functions or may function as a support, in relation to its configuration at the atomic level and to the dynamic interactions among the elements that compose it.

Bibliography

Agassi, J.: Art and Science. In Scientia, **114** (1979) 128–140.

Agazzi, E. (Ed.): La simmetria (memorie del seminario, Venezia, aprile 1970), Il Mulino, Bologna 1973 [Symmetry], only available in Italian.

Arnheim, R.: Entropy and Art. University of California Press 1971.

Arnheim, R.: Visual Thinking. University of California Press 1969.

Bach, J. S.: Preludio, Partita III per violino solo in E major. BWV 1006

Bach, J. S.: The Art of Fugue. BWV 1080

Beretta, G. P.: Punti di vista sul concetto di entropia, Tesi Politecnico di Milano, Facolta di Ingegneria, A. A. 1978–79 [Viewpoints on the Concept of Entropy], only available in Italian.

Borsellino, A., De Marco, A., Allazetta, A., Rinesi, S., Bartolini, B.: Reversal Time Distribution in the Perception of Visual Ambiguous Stimuli. In Kybernetics **10** (1972) 139.

Borsellino, A., Carlini, F., De Marco, A., Penengo, P., Trabucco, A., Riani, M., Tuccio, M. T.: Stochastic Models and Fluctuations in Reversal Time of Ambiguous Figures. In Perception **6** (1977) 645.

Borsellino, A., Carlini, F., Riani, M., Tuccio, M. T., De Marco, A., Penengo, P., Trabucco, A.: Effects of Visual Angle on Perspectives Reversal for Ambiguous Patterns. In Perception **11** (1982) 263.

Bottani, C. E., Caglioti, G., Novelli, A., Ossi, P. M., Rossitto, F., Silva, G.: Il limite elastoplastico: realta fisica e attuali criteri di qualificazione. In Metallurgia Italiana, **7/8**, (1982) 530 [The Elastic-Plastic Limit: Physical Reality and Present Criteria of Qualification], only available in Italian, see also Mechanical Instability of Metals. Europhys. News **14** (12) (1983) 10.

Brillouin, L.: Science and Information Theory. Academic Press, New York 1956.

Brillouin, L.: Scientific Uncertainty and Information. Academic Press, New York 1964.

Butler, J. P., Klug, A.: The Assembly of a Virus. Sci. Amer. **239** (5) (1978) 62.

Caglioti, G.: Simmetria e ambiguita nella fisica e nelle arti figurative. In Annuario dell'Enciclopedia della Scienza e della Tecnica. Mondadori, Milano 1983 [Symmetry and Ambiguity in Physics and in Visual Arts], only available in Italian.

Caglioti, G.: Introduzione alla fisica dei materiali. Zanichelli, Bologna 1974 [Introduction to the Physics of Materials], only available in Italian.

Caglioti, G., Caianiello, E. R.: Ambiguity in Visual Communication: a Case Study. III Symposium International Discoveries, Paris 23–27 oct. 1978.

Caglioti, G., Caianiello, E. R.: A Model for Non-resolvable Ambiguities. In Biol. Cybernetics, **31** (1978) 205.

Caianiello, E.: Some Remarks on Organization and Structure. In Biol. Cybernetics **26** (1977) 151.

Caianiello, E., Di Giulio, E.: Introduzione alla cibernetica. Le Monnier, Firenze 1980 [Introduction to Cybernetics], only available in Italian.

Caillois, R.: Fecondita della dissimmetria. In Agazzi, cit., p. 425 [Fertility of Dissymmetry], only available in Italian.

Calvino, I.: Una pietra sopra, Einaudi, Torino, 1980, [Let Bygones be Bygones] only available in Italian.

D'Amore, B.: Segni, sistemi di segni e strutture. In F. Grignani, "Pitture, sperimentali e graphic design", Comune di Reggio Emilia, Assessorato Istituzioni Culturali, Commissione Arti Figurative e Comunicazioni Visive 1979 [Signs, Systems of Signs and Structures in Paintings, Experimentals and Graphic Design], only available in Italian.

Eco, U.: Il nome della rosa. Bompiani, Milano, 1980 [The Name of the Rose], Secker and Warburg, 1983.

Fast, J. D.: Entropy. MacMillan, London 1970.

Fermi, E.: Notes on Quantum Mechanics, (collected and edited by E. Segré). Phoenix Books, Chicago 1966.

Fermi, E.: Thermodynamics (1936), Italian transl., Termodinamica. Boringhieri, Torino 1958.

Feynman, R. P., Leighton, R. B., Sands, M.: The Feynman Lectures on Physics, vol. 3: Quantum Mechanics. Addison Wesley, Reading 1965.

Frölich, F.: The Linguistic Structure of the Chromosome Genetic Code and Language. In Synergetics, A Workshop, (Ed. H. Haken), Springer Verlag, Berlin 1977.

Gabor, D., Colombo, U.: Oltre l'età dello spreco, Quarto Rapporto al Club di Roma. Mondadori, Milano 1976 [Beyond the Age of Waste], only available in Italian.

Gagliardo, E., Fornasari, P.: Composizione di musica classica mediante elaboratore elettronico. Quaderni di Informatica, **10** (1978) 28–31 [Composition of Classical Music by Computer], only available in Italian.

Gatlin, Lila, L.: Information Theory and the Living System. Columbia University Press, New York 1972.

Glansdorff, P., in "Le Pensée", Revue du Rationalisme moderne, Arts, sciences, philosophie, **195** (1977) 43–62, [Evolution of the Notion Order and Organization in Physics], only available in French.

Glansdorff, P., Prigogine, I.: Thermodynamic Theory of Structure, Stability and Fluctuations. Wiley-Interscience, London 1971.

Grignani, F. A.: Una metodologia della visione. Comune di Milano, Ripartizione Cultura, 1975 [A Methodology of Vision], only available in Italian.

Haken, H. (Ed.): Synergetics. Proceedings of a Symposium on Synergetics, Teubner, Stuttgart 1973.

Haken, H. (Ed.): Cooperative Effects, Progress in Synergetics. North Holland, Amsterdam 1974.

Haken, H. (Ed.): Synergetics – a Workshop. Springer Verlag, Berlin 1977.

Haken, H.: Synergetics, an Introduction. Non-equilibrium Phase Transition and Self-organization in Physics, Chemistry and Biology, 2nd ed. Springer Verlag, Berlin 1983.

Haken, H.: Pattern Formation by Dynamic Systems and Pattern Recognition. Springer Verlag, Berlin 1979.

Haken, H.: Erfolgsgeheimnisse der Natur. Synergetik: die Lehre vom Zusammenwirken. Deutsche Verlag-Anstalt, Stuttgart 1981.

Haken, H. (Ed.): Evolution of Order and Chaos in Physics, Chemistry, and Biology. Springer-Verlag, Berlin 1982.

Haken, H.: Advanced Synergetics. Instability Hierarchies of Self-organizing Systems and Devices. Springer Verlag, Berlin 1983.

Heine, V.: Group Theory in Quantum Mechanics. Pergamon Press, Oxford 1960.

Helmcke, L. G.: Simmetria e ritmo nella natura vivente e nel pensiero umano. In Agazzi, cit. p. 273 [Symmetry and Rhythm in the Living System and Human Thinking], only available in Italian.

Herzberg, G.: Molecular Spectra and Molecular Structure, I – Spectra of Diatomic Molecules. Van Nostrand Reinhold Co., New York 1950.

Koschmieder, E. L., Adv. Chem. Phys. **26** (1974) 177.

Landau, L. D., Lifshitz, E. M.: Physique statistique. Mir, Moscou 1967 [Statistical Physics].

Lee, D. T.: "Symmetry Principles in Physics" in Evolution of Particle Physics, a volume dedicated to Edoardo Amaldi on his sixtieth birthday, (ed. M. Conversi). Academic Press and Periodici Scientifici, Milano 1968.

Lehninger, A. L.: Bioenergetics, the Molecular Basis of Biological Energy Transformations. 2nd. ed. Benjamin, Menlo Park 1971.

Lotka, A. J.: Elements of Mathematical Biology, a Classic Work on the Application of Mathematical Aspects of the Biological and Social Sciences. Dover, New York 1956.

Mathieu, V.: La simmetria in metafisica. In Agazzi, cit. [Symmetry in Metaphysics], only available in Italian.

Monod, J.: Le hazard et la nécessité. [Case and Necessity], available in Italian (Mondadori, Milano 1970) and in French.

Montale, E.: Sulla Poesia. Mondadori, Milano 1976 [On Poetry], only available in Italian.

Morrison, P.: Sulle simmetrie infrante. In "L'estetica nella scienza". Editori Riuniti, Roma 1982 [On Broken Symmetries] in "On Aesthetics in Science", Judith Wichsler ed., MIT, 1978.

Moser, von H.: Messung der wahren spezifischen Wärme von Silber, Nickel, ß-Messing, Quarzkristall und Quarzglas zwischen +50 und 700 °C nach einer verfeinerten Methode. Physikanaliste Zeit. **37** (21) (1936).

Munari, B.: Design e comunicazione visiva. Contributo a una metodologia didattica. Laterza, Bari 1970. [Design and Visual Communication – a Contribution to a New Didactic Methodology], only available in Italian.

Nicolis, G., Prigogine, I.: Self Organization in Non-equilibrium Systems. Wiley, New York 1977.

Pacault, A., Vidal, C. (Ed.): Synergetics, Far from Equilibrium. Springer Verlag, Berlin 1979.

Pantell, R. H., Puthoff, H. E.: Fundamentals of Quantum Electronics. Wiley, New York 1969.

Prigogine, I.: La nuova alleanza: uomo e natura in una scienza unificata. Longanesi, Milano 1979 [The New Alliance: Man and Nature in a Unified Science], available also in Italian.

Prigogine, I., Stengers, I.: La nouvelle alliance. In Scientia **112** (1977) 287–332 and 617–653 [The New Alliance], only available in French.

Prueitt, Melvin, L.: Computer Graphics. Dover, New York 1975.

Radicati di Brozolo, L. A.: Simmetrie e invarianze, in "Enciclopedia del Novecento", Istituto dell'Enciclopedia Italiana, Vol. V, Roma 1982 [Symmetries and Invariances], only available in Italian.

Regge, T.: Cronache dell'universo. Boringhieri, Torino 1981 [Chronicles of the Universe], only available in Italian.

Segre, C.: Semiotica filologica, testo e modelli culturali. Einaudi, Torino 1979 [Philological Semiotics], only available in Italian.

Seneca, L. A.: Lettere a Lucilio, Libro I, Lettera I. Rizzoli, Milano 1966 [Letters to Lucilio, Book I, Letter I], available in Italian.

Shannon, C. E., Weaver, W.: The Mathematical Theory of Communication. University of Illinois Press 1949.

Shannon, C. E., Von Neumann, J., in Bell System Technical Journal, July and Oct. 1948.

Shubnikov, A. V., Koptsik, V. A.: Symmetry in Science and Art. Plenum Press, New York 1974.

Silvestri, M.: Aspetti fisico-matematici della tassazione progressiva sul reddito. Relazione presentata al XXXI Congresso dell'Associazione Termotecnica Italiana, Pavia, sett. 1976. In Termotecnica **31** (1977) 572 [Physico-mathematical Aspects of Graduated Income Taxation], only available in Italian.

Simmetrie, Atti del Simposio, Quaderno n. 136 dell'Accademia Nazionale dei Lincei, Roma 1970 [Symmetries], only available in Italian.

Thom, R.: Structural Stability and Morphogenesis. Benjamin/Cummings, Reading 1975.

Velarde, M. G., Normand, C.: Convection, in Sci. Amer. **243** (1) (1978).

Venkataraman, G.: Fluctuations and Mechanical Relaxation. In Trans. of the Indian Inst. of Metals **32** (6) (1979).

Weyl, H.: Symmetry. Princeton University Press 1952.

Wittgenstein, L.: Tractatus Logicus Philosophicus, Andover, UK, 1964, available in English.

Zorzoli, G. B.: Vivere con il sole, prospettive e limiti delle nuove fonti di energia. Bompiani, Milano 1978 [Living with the Sun, Perspectives and Limits of New Energy Sources], only available in Italian.

MIX
Papier aus verantwortungsvollen Quellen
Paper from responsible sources
FSC® C105338

If you have any concerns about our products,
you can contact us on
ProductSafety@springernature.com

In case Publisher is established outside the EU,
the EU authorized representative is:
Springer Nature Customer Service Center GmbH
Europaplatz 3, 69115 Heidelberg, Germany

Printed by Libri Plureos GmbH
in Hamburg, Germany